Content Strategy 101
Transform Technical Content into a Business Asset

Sarah S. O'Keefe
Alan S. Pringle

SCRIPTORIUM press

What people are saying about Content Strategy 101

Content Strategy 101 is a blueprint for planning and implementing successful content projects.

> –Scott Abel, The Content Wrangler

O'Keefe and Pringle's take on content strategy for technical communicators is a reminder that regardless of our background—marketing or tech, web or print—we need to get out of our silos and work together toward our one real goal: helping our customers.

> –Sara Wachter-Boettcher, content strategist and author of *Content Everywhere*

Finally, a book on content strategy for "the rest of us." O'Keefe and Pringle fill a need with this book. All of us in tech comm have things to learn from them.

> –Marcia Riefer Johnston, author of *Word Up!*

Sarah and Alan share from their many years of experience helping executives manage their technical documentation process more efficiently to improve their bottom line.

> –Tony Chung, Swiss-army knife for the web

Want to provide users with great content while simultaneously streamlining your processes and costs? Then *Content Strategy 101* by Sarah O'Keefe and Alan Pringle is a must read.

> –Jacquie Samuels, Information Architect

Finally! A plain language, easy-to-understand book about content strategy for technical communicators. Written in a frank, conversational style, the book illustrates, beyond a doubt, all the reasons that organizations will benefit from implementing a content strategy, or at least seriously considering it.

–Rahel Anne Bailie, Senior Content Strategist,
Intentional Design

As experienced practitioners, O'Keefe and Pringle have performed a real service to our industry by providing a very practical guide that will ultimately lead people to come to grips with the real content within their organizations. To my mind, too much advice circulating on content strategy in fact has very little to do with content and even less to do with strategy. With solid handle on practicalities like implementation costs, tangible savings and revenue opportunities, and a clear view to where the technical content in question actually comes from, this book is a useful corrective.

–Joe Gollner, Director, Gnostyx Research

Content Strategy 101 resounds well as a nice resource for the technical communicator who is looking to make waves within their business or department.

–Roger Renteria, TechWhirl writer and author of
writetechie.com

Content Strategy 101 is a great book for those managers who know that great content is important and that there must be a way to produce it in an efficient way, but aren't quite sure where to start. I'd recommend it for anyone: those managing the process and those attempting to deliver the works.

–Al Martine, Partner, TechWhirl.com

Contents

Contents

About the authors

Sarah O'Keefe is the chocoholic founder of Scriptorium Publishing (scriptorium.com) and a content strategy consultant. Sarah's focus is how to use technical content to solve business problems; she is especially interested in how new technologies can streamline publishing workflows to achieve strategic goals.

Sarah speaks fluent German, is a voracious reader, and enjoys swimming, kayaking, and other water sports along with knitting and college basketball. She has strong aversions to raw tomatoes, eggplant, and checked baggage.

You can follow Sarah on Twitter at sarahokeefe.

Alan Pringle has worked for Scriptorium Publishing since its inception in 1997. As director of publishing operations, he helps clients solve problems with the development and distribution of technical content. His responsibilities include content strategy analysis, automating production and localization tasks (often with XML-based workflows), and managing schedules and budgets.

A native North Carolinian, Alan enjoys eating (preferably pastry or chocolate), reading, traveling, and watching vintage bad movies.

You can follow Alan on Twitter at alanpringle.

Acknowledgments

This book builds on the work of industry leaders, especially:

- *Managing Enterprise Content: A Unified Content Strategy* by Ann Rockley and Charles Cooper
- *Content Strategy for the Web* by Kristina Halvorson
- The Content Wrangler web site, Scott Abel

Portions of this book are based on blog posts first published on our web site (scriptorium.com/blog). Numerous people provided comments on those posts and on early drafts of this book at contentstrategy101.com, including:

Andrea Wenger	Marijana Prusina
Axel Regnet	Mark Baker
Denise Kadilak	Michael Müller-Hillebrand
Erin Vang	Nicky Bleiel
Frank Buffum	Pamela Clark
Larry Kunz	Rick Sapir
Jennifer O'Neill	Tamsin Douglas
Kai Weber	

Special thanks to our early readers, Larry Kunz and Marcia Riefer Johnston, who caught some really embarrassing typos, sloppy arguments, and generally lazy writing. The remaining howlers are of course the authors' responsibility.

Gretyl Kinsey created most of the graphics. Simon Bate designed the DITA Open Toolkit PDF plugin for this book. Ryan Fulcher and Holly Mabry developed the companion contentstrategy101.com site.

Dick Johnson of VR Communications provided the missing link—an automated way to export the DITA source files over to WordPress. This bit of programmer magic allowed us to post several drafts and reap the benefits of the community's comments.

The information in "A historical perspective on content" was previously published in several versions:

- "What Do Movable Type and XML Have in Common?" December 2008, *Intercom* magazine (scriptorium.com/movabletypexml.pdf)

- "The economics of information," August 2011, *tcworld* magazine (tcworld.info/tcworld/content-strategies/article/the-economics-of-information).

Foreword

As companies work to define more effective ways of reaching and engaging their customers, the topic of content strategy often comes up. All too often, content strategy is applied only to the most visible information: marketing collateral and web sites. Useful as they are, these are not the only types of content that can benefit from a rigorous application of content strategy. The information created by technical communicators—*documentation*—is just as important, but for a long time has received short shrift when compared to its more high-profile brethren.

Technical content is often thought of as a "necessary evil," but in fact good documentation is critical in supporting the product-buying decision and in reducing the cost of product support.

Note that I said *good* documentation. Good documentation is a valuable part of a product, but without an effective content strategy, your chances of getting that "good documentation" are about as good as the chance of a snowball in… well, you get the idea. You need a content strategy for *all* your content—but where to start?

Content Strategy 101 is the perfect place.

Content Strategy 101 helps managers to understand the key factors required to build a business case, and the book guides practitioners through the many aspects of a content strategy they need to investigate and implement.

Sarah O'Keefe and Alan Pringle draw on their long-term experience in the technical content industry to cover a broad range of issues affecting a good content strategy, including business processes, solution architecture, the creation of good content, and the role of technology. And they don't just talk about traditional print and help-based content; they discuss the role of web-based, wiki-based, and customer-generated content.

Sarah and Alan's *Content Strategy 101* is a much-needed book that fills a significant gap in the discussion of content strategy.

Ann Rockley

Preface

In 1997, we established Scriptorium Publishing with a mission to streamline publishing workflows. Over time, the tools and technologies have changed, but not our focus on automation and efficiency. In the mid-2000s, the term "content strategy" was first applied to our work.

Today, publishing is being revolutionized by two factors:

- The rise of digital content (such as ebooks) that eliminates distribution costs and increases publishing velocity

- The rise of community content (blogs and wikis, for example) that disregards content gatekeepers

Many technical content producers are ill-equipped for these changes. They cling to the old way of doing things because it's comfortable and familiar. But their printed books (and PDF files) are competing with third-party blogs—and Google search performance determines the winner.

Content Strategy 101 gives you a roadmap for understanding your business content requirements. This book will help you understand the different content options, identify the best choices for your unique requirements, and develop a strategy for your technical content.

Most of the books written about content strategy focus on web content strategy, which is usually synonymous with marketing content strategy. *Content Strategy 101* focuses on the often-overlooked technical content. We assume that you are responsible for technical communication in your organization—traditionally, user manuals and online help. You probably already make this information available on the web or are trying to do so. You may also create other information products[1], such as podcasts, videos, screencasts, technical illustrations, and posters. If you are trying to establish a content strategy for these core technical documents, this book is for you.

Chapter 1, "Getting started," discusses the history of content and publishing before turning to the specifics of technical content and the modern concept of structured, intelligent content.

After that, the book is divided into three parts:

- **Part I: Business goals** describes how technical content can contribute to controlling technical communication costs, improving product marketing, and ensuring legal and regulatory compliance.

- **Part II: Developing a technical content strategy** explains how to turn the business goals you identified into a plan.

- **Part III: Implementing your content strategy** provides a method for going from the plan to reality.

We hope that you find this information useful.

Please visit the companion web site, contentstrategy101.com, to join the conversation.

---

[1] For lack of a better term, "information product" means any content collection, including books, PDF files, ebooks, web sites or pages, and help systems. The word "document" is simpler but has print connotations. Other than in academia, "text" does not include audio, video, or graphic content.

Chapter 1: Getting started

It used to be so simple. A technical writer would meet with an engineer, gather information, write it up—in longhand—on a legal pad, and then send the information off to the typing pool. After some revisions, the typed manuscript and perhaps hand-drawn graphics would be delivered to the printer and, eventually, a book appeared. Over time, the legal pads were replaced with typewriters; then, the typewriters were replaced with computers. In addition to producing text, technical writers accepted responsibility for page layout and pre-press production tasks.

Today, technical writers are more often technical *communicators*: they produce text, images, photographs, charts, live video, screencasts, webcasts, comic books, simulations, and more. And technical communicators face a bewildering array of options: XML, help authoring tools, wikis, customer-generated content, desktop publishing tools, conversion tools, and so on. Instead of creating content in isolation, technical writers coexist with training, collaborate with technical support, and compete with user-generated content.

Other factors further increase the complexity:

- *Global markets require global content.* You must create information in your customers' languages or, as a fallback, simplify content so that readers with limited proficiency in the language provided can understand it.

- *Product development cycles are shorter.* Information needs to be updated more often. A document production process that takes a week per iteration is perhaps acceptable for yearly product releases, but not for a quarterly update schedule.

- *Government regulations and compliance requirements have increased.* Regulatory agencies mandate not just what information needs to be delivered but the storage format of that information.

- *Product variants or custom products are more common.* Buyers expect customized content that reflects their unique configuration.

Writers cannot "just write." First, they must decide what information is needed to support a particular product, who will create that information, and how best to deliver the information to their audience. They need a *content strategy.*

The best-known definition of content strategy comes from Kristina Halvorson of Brain Traffic, who says that content strategy:

> plans for the creation, publication, and governance of useful, usable content. [2]

Rahel Bailie of Intentional Design uses this definition:

> A repeatable system that governs the management of content throughout the entire lifecycle. [3]

Content Strategy 101 is a call to action for anyone involved with technical content: writers, managers, and executives. Without a content strategy, you will waste time and money with inefficient processes to create information products that do not support your business goals. Scott Abel, The Content Wrangler, says that "Content is a business asset worthy of being managed." When done poorly, technical content is a liability—it can result in damage to your reputation, lost sales, and legal problems.

You need a content strategy to ensure that you:

- Deliver the right information

- Deliver information effectively

---

[2] "The Discipline of Content Strategy," A List Apart web site, published December 16, 2008, alistapart.com/articles/thedisciplineofcontentstrategy/, visited August 7, 2012

[3] "What's the buzz about content strategy?," *tcworld magazine*, August 2011, tcworld.info/tcworld/content-strategies/article/whats-the-buzz-about-content-strategy/, visited August 7, 2012

- Engage your customers and build community
- Streamline your publishing process
- Meet legal and regulatory requirements
- Control the cost of content

A historical perspective on content

In Europe before the 1450s, books were precious, rare objects and were usually copied by hand over a period of months or years. Johannes Gutenberg and his printing press changed the economics of information distribution. The result of this change was less expensive books, greater literacy, and a challenge to those in power, who benefited from restricting information. Today, the rise of the Internet has eliminated distribution costs as a barrier to entering the publishing market. With minimal equipment, anyone can publish in a blog or book, record and distribute a podcast, or deliver video content. What do these changes mean for technical communication? And what lesson can we learn from the changes that took place over 400 years ago?

In the last 20 years, the economics of information have shifted toward the author and away from the publishers (or gatekeepers):

- It's possible to record high-quality audio and video with inexpensive equipment
- The Internet provides numerous publishing platforms (Blogger, WordPress, YouTube, Lulu, Amazon, iTunes, and so on)

The possibilities are endless: books, ebooks, PDF files, web content, screencasts, podcasts, digital videos, wikis, and more. But which of these platforms will succeed?

The text cycle

To understand the implications of digital publishing, it's helpful to break down the process of information development. Terje Hillesund developed a text cycle[4] with the following phases:

- Writing (authoring)
- Production
- Storing
- Representation
- Distribution
- Reading (consumption)

Traditional storytelling combines all of these phases into a single event: one person at the campfire telling a story while the audience listens.

The written language separates distribution and consumption. Instead of needing an author to deliver the story in person, written content can be moved from one location to another.

The printing press introduces further separation of the phases by disconnecting production (formerly hand-copying) from distribution. It becomes possible to produce a page once and create many, many copies of that page.

Digital content allows further separation. Physical distribution is no longer required, and the representation (formatting) of the text is separated from the production (markup) and potentially from the storing (content management system).

Quality versus cost

The printing press, which made inexpensive books possible, did require a compromise in quality. Hand-crafted, hand-copied books, with their carpet pages, intricate capital letters, and unique illustrations (often customized for the person who commissioned the book) were works of art.

---

[4] In this context, "text" includes graphics and other content types.

The earliest printed books were hand-illuminated after the printing process, but this added effort gave way quickly to mass-produced books. The ability to produce books faster and cheaper was more compelling than the increased quality resulting from extra manual work.

Before the printing press, the act of copying the book also created the formatting. With the printing press, the formatting was done in a separate typesetting step, and it was then possible to create a large number of copies from a single formatting effort.

Today, the publishing world sits at a very similar inflection point. The rise of electronic publishing along with the ability to separate authoring from formatting is analogous to the rise of printing and the ability to separate formatting from distribution. Just before the printing press (1450), approximately 50,000 books existed in Europe. Within 50 years, that number rose to 12 million.[5]

Figure 1: The rise of books in Europe after the printing press

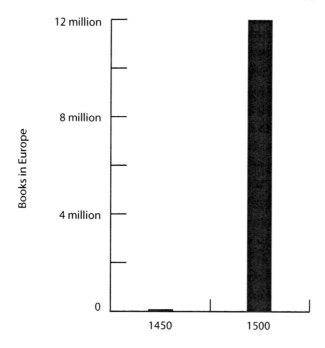

---

[5] hrc.utexas.edu/educator/modules/gutenberg/books/legacy/

What are the implications for technical content?

The rules of publishing, which were relatively static for hundreds of years, are now changing by the day. Consider that iPad tablet publishing did not exist until 2010. The Kindle reader (2008) drives a brand new ebook business. We can expect to see increases in publishing velocity, volume, and versioning requirements. And based on the way that printing evolved, we can expect that economic considerations will determine which innovations succeed.

With this in mind, expect the following developments:

Streamlined publishing workflows

Given the proliferation of output formats, the publishing workflow must be automated. Labor-intensive final production work will likely disappear. Like hand-illumination, these tasks add quality, but they obstruct efficiency. For technical content, efficiency will outweigh perfect kerning, copy-fitting, and other design niceties.

Data-driven, user-customizable graphics

Highly designed infographics and other complex images will remain the domain of the professional author for now. To reduce the cost of maintaining (and especially translating) these graphics, authors must use layers and carefully separate the core graphic elements from the labels that require translation. There is room, however, for growth in graphics that readers can manipulate or create. If we make the data available to our readers, they can choose how to display the information (bar graph or pie chart?), filter the information displayed on the chart, and control the colors and the fonts used in the chart. Google Analytics and many web-based application dashboards provide users with ways to manipulate data. Technical communication needs to make better use of these types of technologies and provide flexible ways to render information. Instead of focusing on controlling the presentation of graphical information, we can build information applications that the reader can control.

Limited use of audio and video

If we apply Hillesund's text cycle to audio and video, we can see why audio and video are not (yet) going to take over from text. The components of the audio and video development cycles are not yet separated as clearly as the text development components. In particular, when audio or video is recorded, the content storage and representation are tied together. These two facets need to be separated to provide for really inexpensive (and therefore widespread) usage. A basic example where storage and representation are separated is text-to-speech functionality, which has the ability to render audio in a voice chosen by the end user, rather than in the audio track laid down by the author. But the vast majority of audio files use sound recordings, where the content is inextricably tied together with the delivery. There are similar issues with video. Exceptions are screencasts and digital animation, where the source files have layers and timelines, which content creators can manipulate as needed. But today, we do not have the same degree of separation of content and formatting for audio and video as we do in text and graphics. We can't slice apart audio and video the same way that we manipulate text.

Velocity, volume, and versioning

Velocity, volume, and versioning are the three Vs that drive the economics of information:

- *Velocity:* the speed at which new information is created and delivered

- *Volume:* the amount of content that needs to be created and delivered

- *Versioning:* the content variations that need to be supported for end users

The requirements for the three Vs are pushing organizations to fully automate their workflows to eliminate delays in information delivery.

Velocity and volume are also implicated in the rise of topic-based authoring. When authors work at the topic level, it's easier to move authors from project to project and therefore put additional people to work on high-priority projects. This is much more difficult in narrative or book-based content.

Like velocity and volume, versioning requirements are increasing. Instead of creating a few manageable versions of content, technical communicators are being asked to support products that have dozens or hundreds of variations. The only reasonable solution with the higher number of versions is to deliver all of the content, and then filter it based on a user's profile. This requires an excellent understanding of the product and (again) complete automation of the rendering process.

High-end versioning probably means that the content objects need traceability—they need to be connected to the corresponding product functions, so that the system can include the appropriate information for each user.

It's worth noting (again) that the three Vs apply mainly to text and somewhat for graphics.

Search and navigation

Information is valuable only if users can access it. For books, we have standard conventions: a table of contents at the beginning of the book, an index at the end, and page numbers for navigation. We also know that a book in English is read from left to right. Chapter title pages, headings, and caption for figures and tables are all instantly recognizable because we have been exposed to them since primary school.

For newer electronic information products, search and navigation are even more critical—"flipping through the book" is really not viable online—but the user experience is not yet unified. The behavior of an EPUB file depends on the capabilities of the reader in which it is being displayed.

Figure 2: Content displayed in the iBooks app on an iPad tablet

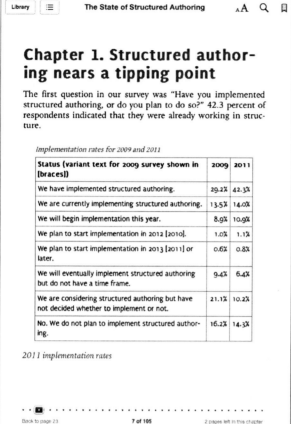

The same EPUB file renders differently on a NOOK reader.

Figure 3: Content displayed on a NOOK reader

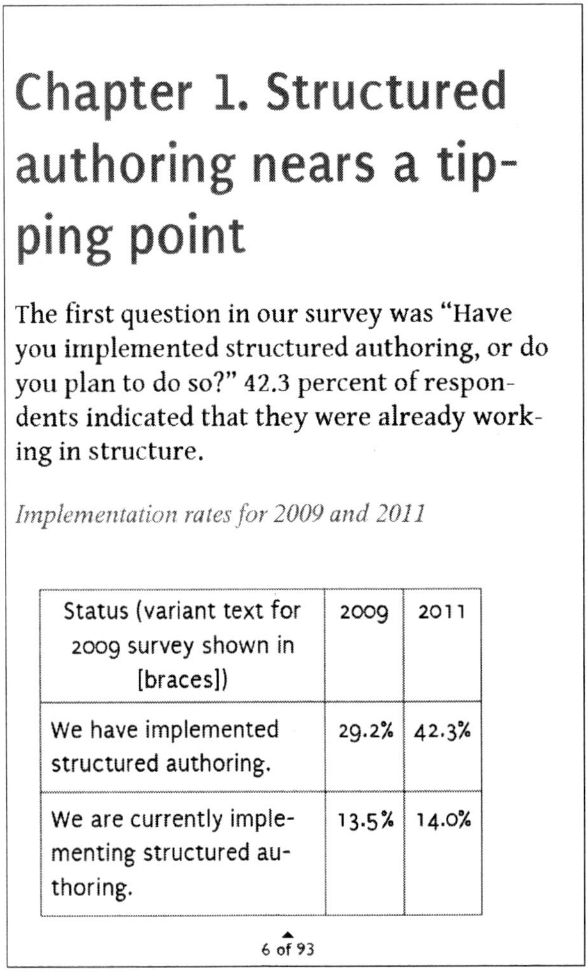

Chapter 1. Structured authoring nears a tipping point

The first question in our survey was "Have you implemented structured authoring, or do you plan to do so?" 42.3 percent of respondents indicated that they were already working in structure.

Implementation rates for 2009 and 2011

Status (variant text for 2009 survey shown in [braces])	2009	2011
We have implemented structured authoring.	29.2%	42.3%
We are currently implementing structured authoring.	13.5%	14.0%

6 of 93

Adobe Digital Editions provides yet another variation.

Figure 4: Content displayed in Adobe Digital Editions

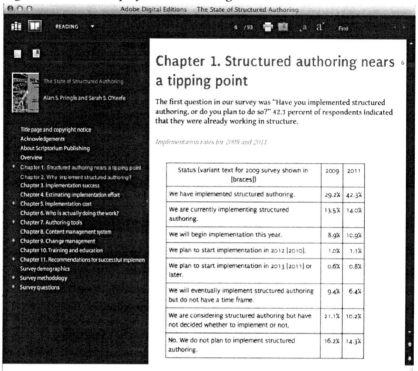

For content creators, this introduces a lot of headaches. For example, consider an interactive, multimedia-rich ebook. In this context, what is the equivalent of a page number? What if the rendering and the page numbers are different on different devices?

In addition to navigational issues, there are search challenges.

How will users find the information that they need? Search provides a partial answer, but even the most carefully crafted search string may result in an overwhelming list of results. Search with filters (faceted search) and social search (results that are influenced by the searcher's social network) can make results more manageable.

Figure 5: Faceted search (left) and social search (right)

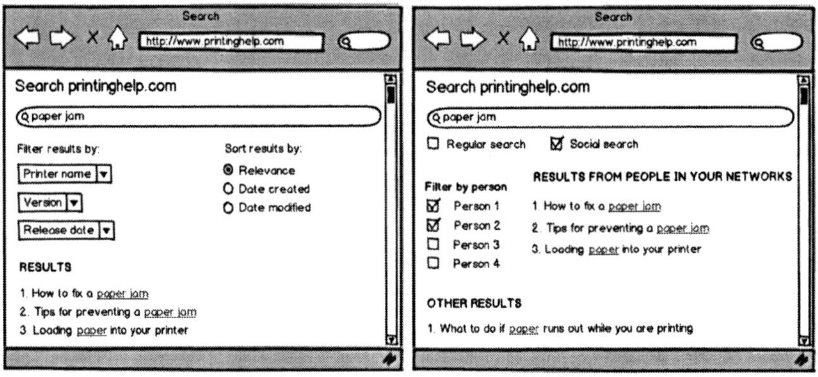

The world of technical content

The functional goal of technical content is to help people use a product successfully. The business goal must tie the content into organizational strategy. For instance, in a regulated industry, creating required documentation is a prerequisite to entering the market.

Technical content may have persuasive objectives; for example, describing the additional features available in a more expensive product edition and thereby encouraging people to upgrade. But the informational angle is more important than the persuasive function. Persuasive content is marketing content, where the primary goal is to convince the reader to buy a product or service. This difference in perspective separates the practices of technical communication (tech comm) and marketing communication (marcom).

Perhaps Tim O'Reilly, founder of O'Reilly Media, says it best. Technical information is "changing the world by spreading the knowledge of innovators."

But not all technical content contributes to this mission. Most everyone has attempted to decipher a manual or support web site that uses incoherent language, atrocious organization, and appalling formatting. These information products are unusable because the responsible organization does not believe that technical content quality affects business performance. But is this true?

When managed properly, technical information can help you to:

- Meet regulatory requirements with minimum cost
- Extend your global reach by delivering content optimized for each market
- Reduce technical support costs
- Reduce product returns
- Improve customer satisfaction
- Lower the overall cost of information development

Historically, tech comm has been a cottage industry where each technical writer is responsible for a specific content area. Today, tech comm is moving into a manufacturing model. This approach requires a huge shift in mindset for experienced writers. Instead of owning all aspects of a book or help system, writers become content contributors who collaborate to produce a final product. This approach has advantages for the organization, especially for larger writing groups, but it is highly disruptive to the traditional approach because of the following factors:

- In traditional tech comm, a writer is responsible for a defined information product, usually from start to finish. That means developing an outline, creating the content, and formatting it for the final delivery (such as PDF or HTML). In a collaborative approach, a writer might have an assignment to write just 25 percent of a document and have no input on the final document presentation. It is the difference between hand-crafting an automobile and working on a specific component on an assembly line.[6]

- Writing in a collaborative environment is much more difficult than writing solo. The information must be consistent so that content from different writers can be assembled into a coherent document. This requires strict adherence to style guidelines and avoiding unique turns of phrase that stand out from the larger document.

---

[6] For a detailed exploration of this idea: "You can have any color, as long as it's black," Dr. Tony Self, first published in *ISTC Communicator,* Spring 2012, (hyperwrite.com/Articles/showarticle.aspx?id=90).

Most managers and executives treat technical communication as a cost center. *Content Strategy 101* focuses on how to evolve technical information from "necessary evil" to "business asset." Some examples of this transformation include the following:

- Improve content by removing extraneous information ("minimalism"), which results in reduced development and localization costs.

- Improve templates and formatting automation, thus reducing manual formatting time in each product release.

- Use search engine optimization techniques to improve content's Google performance, increasing page views and therefore visibility of your content. (Then, show a correlation between increased web site traffic and reduced technical support calls.)

- Ensure that information is shared across the organization, so that different groups do not waste time writing the same content twice.

For technical communicators and consultants, a stronger focus on business value means that they must:

- Have a reasonable level of writing competence. In most cases, an adequate writer who produces content quickly will be preferred over a great writer who works more slowly.

- Produce content that conforms to established style guidelines, templates, and other corporate standards. (Clean content is more valuable than messy content.)

- Look for ways to improve productivity and drive down the cost of content development infrastructure.

- Focus more on efficient technical communication and less on the art of writing. For example, they might develop innovative new ways to create and deliver content, such as increased community participation.

Competing priorities

Some technical writers see themselves as user advocates—they identify with the user who needs help using a complex product. The "user advocate" perspective needs to be balanced with the business priorities.

Technical communicators are paid not by the user, but by the organization, and that means focusing on business value and not purely on users. The most egregious offenders wield "the user" as a club to justify their personal style and content preferences.

The technical communication group is not the ombudsman for the user. Close collaboration with product developers is needed instead of an antagonistic relationship based on the product developers' perceived lack of empathy for the end user.

A successful content strategy for technical content must balance the following, often conflicting, issues:

- *Strategic.* Technical content must support business goals. If certain information is required in the documentation to meet legal requirements, it must be included even if it is not useful to the reader. If the goal of technical content is to support marketing, some persuasive information is appropriate.

- *Useful.* Technical content must provide readers with the information that they need in an appropriate format.

- *Efficient.* The process of creating technical content should use limited resources wisely.

Figure 6: Balancing competing priorities is essential for successful technical communication

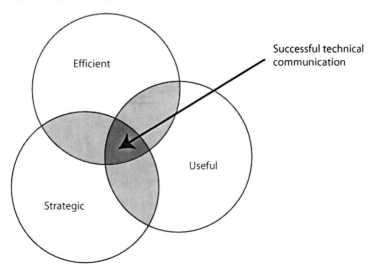

For a summary of the components that contribute to good technical content, see Appendix A, "Creating useful information."

An overview of content delivery options

Here is a quick overview of some of the most common formats used to deliver technical content.[7]

Paper

The only option until the 20th century provides a huge variety of items ranging from art books ("coffee table books") to laminated quick reference cards. In technical communication, printed information products are increasingly rare because of cost considerations. Paper is used mainly in locations where electronic deliverables are not technically feasible, such as mines, deserts, and other remote locations, or disallowed by policy, such as many retail organizations and secure facilities (government, utilities, and others). Books are perceived as having high value, so including an attractive book with a product helps the positioning.

Portable Document Format (PDF)

PDF files are an electronic substitute for paper documents. A recipient of a PDF file can print the file and get a fully formatted document that matches the paper that used to be shipped. This is the most common use of PDF files—as a substitute for shipping paper documents. Instead, the cost of printing is shifted onto the recipient, and the time and expense of distributing paper copies are eliminated. PDF files provide some enhancements over paper books, including full-text search, support for 3D images and animations, and live links. PDF files, however, tend to be large and hard to manage.

Web sites

Web sites are another broad category that encapsulates a huge range of content. For technical content, the most common approach is to offer a dedicated site for the documentation (often with a subdomain like docs.example.com) and to provide search capability for the content. Many organizations use the web for content

---

[7] "Key Trends in Software User Assistance," Joe Welinske, 2012, writersua.com/articles/keytrends/index.html

distribution but provide only PDF files and not HTML versions of their technical content. Web sites can be available to the general public or restricted to users with varying levels of credentials required. Only sites that are available to the public are indexed by public search engines.

Multimedia tutorials

Multimedia tutorials are short videos or screencasts (animated screen shots that show a user working in a software application), usually with a voiceover or captions. Tutorials are useful for explaining complex tasks (such as machine repair) and complex software interfaces.

Ebook formats

The ebook landscape is still in flux. The two important formats are currently EPUB and mobi. EPUB is a standard storage format maintained by the International Digital Publishing Forum (IDPF).[8] The iOS iBooks reader and Barnes & Noble Nook support EPUB files. The other format, mobi, is used by Amazon Kindle devices and other ereaders. Amazon's proprietary AZW format is a compressed mobi file, and KF8 format (also proprietary to Amazon) includes mobi and EPUB in one file. In general, technical ebooks are seen as a replacement for print or PDF files. An "enhanced" ebook could go beyond a basic paper replacement and offer interactivity to take advantage of the ereader's capabilities.

Online help

This category includes compiled help (such as Microsoft HTML Help) and "web help," a general term for help that is presented in a web browser (rather than a dedicated help viewer). Typically, web help includes a navigation frame on the left for table of contents, index, and search (often called "tripane help") and a content frame on the right for information. The tripane approach, however, is rapidly going out of fashion because of problems on mobile devices. The term "online help" is showing its age—it refers to help on a computer (as opposed to on paper) and not necessarily to help that is on the Internet. Online help can be embedded with software, shipped on a CD, or delivered over the web.

---

[8] idpf.org/epub

Knowledge base

Knowledge bases are most often used by technical support organizations. They are intended for articles, such as explanations of unique configuration problems, that are created by technical support staff as they provide phone- or email-based assistance.

Forums

Discussion forums and bulletin boards predate the Internet. The conversation can be raucous at times; for corporate forums, it is critical to establish and enforce ground rules. That said, a heavy-handed moderator will drive participants away to less-restricted venues. Too little moderation will drive away participants who do not want to wade through personal attacks and irrelevant political postings to find the information they need.

Wikis

A wiki is a web site whose content can be edited collaboratively by its users.[9] The most famous example is Wikipedia (wikipedia.org). Many organizations provide advanced technical content in wikis and let customers comment on it or edit it. Wikis are most appropriate for detailed content that goes beyond the basics and changes rapidly. Examples might include code examples for a programming language. Wikis are used heavily in user-generated references for online games.

Modern content: structured and intelligent

Structured content is information that is organized in a predictable way. *Structured authoring* refers to creating content in an environment where the required structure is enforced by the authoring software. Most traditional authoring tools, such as word processors and page layout tools, provide a way to define content organization (for example, through templates) but do not enforce the organization (you can choose not to follow the templates). In a structured authoring workflow, following the template is not optional.[10] Structured authoring is often paired with *modular writing*; authors develop content in chunks that are

---

[9] Source: Google search for "define:wiki," accessed August 13, 2012

[10] For more details, refer to "Structured authoring and XML" (scriptorium.com/2009/12/structured-authoring-and-xml/)

assembled to create manuals, web content, help systems, data sheets, and other information products.

Intelligent content is "content that is structurally rich and semantically categorized, and is therefore automatically discoverable, reusable, reconfigurable, and adaptable," according to Ann Rockley and Charles Cooper, authors of *Managing Enterprise Content: A Unified Content Strategy.*

Not every content development effort requires intelligent content, and establishing a structured authoring environment demands a significant investment of time and resources. But to meet the more complex challenges of technical content, a strategy built on structured, intelligent content will allow you to:

- Separate content from presentation so that you can easily deliver information in a variety of output formats

- Tailor content to specific requirements (such as user level, access permissions, installation profile, and more)

- Ensure that users' content searches are successful

As delivery requirements grow increasingly complex with more and more devices (smartphones, tablets, computers, ereaders), multiple languages, and multiple audiences, intelligent content provides a solid foundation for your content strategy.

Part I
Business goals

Chapter 2: Controlling technical communication costs

Technical content is often the last in line for investment and innovation, but poor content has profound effects inside and outside the organization. Before relegating technical content to the "necessary evil" role with minimal investment, consider whether it might actually be less expensive to create high-quality technical information.

The issues to consider are:

- The real cost of low-cost documentation

- How to create an efficient content development process

- Whether high-quality documentation can lower the cost of technical support

- The most cost-effective way to share technical content across the enterprise

The fallacy of low-cost documentation

In many organizations, management pays as little attention as possible to content. Content development is assigned to administrative staff using whatever tools are lying around (usually Microsoft Office) or outsourced to the lowest bidder.

This approach does have advantages:

- *Minimal resources.* Management ignores content and focuses on other priorities.

'he content budget looks small.

ent of this approach usually reveals a number of

...ul. ("Nobody reads the documentation.")

Content is available only in a single output format, such as PDF or perhaps the original Word files.

- Content creators don't understand the product, so they are producing superficial documentation guaranteed to infuriate anyone who does read it. ("In the Name field, type the person's name.")

It is acceptable to assess your organization's content requirements and embark on a strategy of producing indifferent content cheaply (the "meh" strategy). The vast majority of organizations who adopt a laissez-faire attitude, however, have not thought through the implications.

The arguments for a strategy of indifference are:

- Our content is not important to ensure safe operation of our product.
- Our content meets regulatory standards, if any, and customer requirements.
- Providing better content will not help the business.
- Providing content in more formats will not help the business.
- We do not localize content and will never be required to do so.

The argument usually breaks down in the latter portion of this list. Here are some typical business problems that have bad documentation as their root cause:

- *Call volume to technical support is high.* Customers give up on looking in the (terrible) documentation and call instead, thus shifting the cost of bad documentation from the customer back to your organization.

- *Product returns are high and sales are lost.* Customers fail at installing and configuring the product, so they return it. The installation and configuration documentation is impenetrable. There's even a blog warning people not to buy the product because of these problems.

- *Missing content.* A regulatory submission is delayed or rejected because it does not conform to agency requirements. A critical section was left out inadvertently.

- *Cannot deliver required formats.* Use of a military standard for content is specified in a defense contract. But content was developed in a word processor. Now, the profitability and timely delivery on the contract are jeopardized by unexpected content conversion.

- *Ugly content contradicts premium product messaging.* The organization markets a product as a high-end product at a premium price. But the documentation looks terrible, which contradicts the marketing message. The manuals do not contribute positively to the initial out-of-the-box experience. Things don't get any better when customers look inside the manuals, either.

- *Huge globalization costs.* The organization has identified opportunities in global markets, but delivering localized content from the existing workflow is unsustainable—the organization isn't even keeping up with the content in the primary language.

- *Technical support and other internal organizations are creating content that duplicates documentation.* Technical support, training, and other workgroups need the content in the documentation, but they are unable to find it. The manuals are delivered as monolithic Word or PDF files, and searching those files is tedious and time-consuming. Instead, the support group resorts to re-creating content in an unofficial knowledge base.

Ignoring content can have huge cost implications across the organization.

Efficient technical content development

The process of creating technical information includes writing text, creating graphics, recording audio, and the like. A basic prerequisite for good content strategy is that the content should be of good quality—accurate, concise, and complete.

An efficient workflow with professional technical communicators creating high-value information is typically *the least expensive* option (better, faster, *and* cheaper). This correlation is explained by the following factors:

- *Reuse versus copy and paste.* Copying and pasting is quick and easy initially, but it is hard to maintain over time because of information duplication. Formal reuse, where a linked copy of information appears in multiple places, is easier to maintain. Given several thousand pages of content, the savings on content maintenance add up quickly.

- *Formatting.* "Quick-and-dirty" formatting is also unmaintainable. Investing in content standards, such as templates, means that content creators spend more time writing and less time formatting. Better tools also mean less time wrestling with pagination, tables of contents, and similar components.

- *Production.* Instead of converting content from one format to another manually, a professional workflow generates the proper outputs automatically. A one-time configuration effort replaces the repeated conversion task.

- *Localization.* Better-organized content and automated production result in much lower localization costs. The "cheap way"—slamming everything into a word processor and throwing it over the transom to the translation vendor—results in escalating translation costs because content must be reformatted in every language. Also, translation costs skyrocket when source files for content are duplicated to accommodate even slight variations in similar products; systematic reuse of content eliminates the expense of translating shared content again and again.

A rough estimate is that writers spend around 20 percent of their time on formatting tasks in less-efficient content development environments (although we have seen much higher numbers).

For mature audiences only

One critical aspect of content development is to understand the maturity level of the process used to develop the content:

1. *Crap on a page.* There is no consistency in content. For example, two white papers from the same company are formatted inconsistently, are often badly written, and do not use consistent terminology. Two audio files might be encoded differently or have wildly varying levels of audio quality.

2. *Design consistency.* Content appearance is consistent, but the methods used to achieve the look and feel vary. For example, two HTML files might render the same way in a browser, but one uses a CSS file and the other uses local overrides.

3. *Template-based content.* Content appearance is consistent, and the methods used to achieve the look and feel are consistent. For example, all HTML files use a common CSS file, or page layout files use the same formatting template. Graphics are created, scaled, and rendered the same way.

4. *Structured content.* Content is validated against a template by the software. This usually means that XML is the underlying file format. Information is organized in predictable, consistent ways.

5. *Content in a database.* Information is stored in a database and can be searched and manipulated in interesting ways.

The bigger the gap between the current authoring environment and the environment needed to implement your strategy, the more difficult the transition will be. For example, if your content strategy requires structured content (level 4), and authors are currently producing whatever they want (level 1), expect significant training challenges, change resistance, and probably staff turnover.

A high maturity level for your process is the means, not the end. The goal of your content development efforts is to support your business goals. A mature process, indicated by template-based content, structured content, or database publishing (levels 3–5), can support business goals because of the following:

- Consistency in development and delivery results in a consistent user experience for the content consumer, which increases their trust in the product. Put another way, if the documentation looks terrible for a premium product, the appearance of the documentation undercuts the premium positioning.

- Accelerating the content delivery process means that you can provide information to end users faster, which makes it more relevant to the end user.

- Formatting is automated, which reduces the overall cost of developing content. This effect is multiplied for each delivery medium (PDF, HTML, and the like) and each language.

For example, a less mature, ad hoc approach (typically, a collection of messy Word files) may be feasible for 500 pages delivered in PDF and HTML in two languages. But once you have more than 2,000 pages, add another output format, or need more than two languages, you need to pay attention to your process and your content quality to ensure that you can scale up to this level—or expect costs to balloon and quality to decrease.

Figure 7: For scalability, you need efficient workflows and high-value content

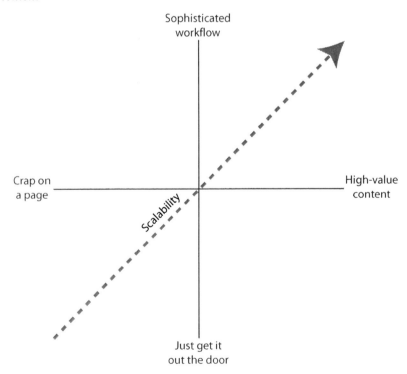

This book concerns itself mainly with the upper right quadrant—creating high-value technical content efficiently.

The right tools and technologies will make your content development process more efficient. (The less pleasant corollary: The wrong tools, even when they are "free," can be very expensive.) An efficient process is one where the tools and technologies that are in place help you produce the content you need, in the formats you need, and when you need it.

Reducing the cost of technical support

Live technical support, whether provided via phone, email, or chat, is expensive because it is provided one-to-one. Technical content, by contrast, is one-to-many and less expensive per person—assuming that you have a reasonable number of readers. Many people can use a single installation guide without increasing the cost of the document. Therefore, many large companies measure the effectiveness of improved content by looking for *call deflection*—a reduction of technical support incidents.

Call deflection isn't the only way in which technical documentation can reduce support costs. If technical content is both reliable and easily searchable, support agents can quickly find the specific information a customer needs—and shorten the average time per call.

Better technical information, fewer support calls

Better content results in fewer calls because users find the information before picking up the phone. You can evaluate the cost of producing and distributing technical content against the cost of providing technical support. For large organizations, which may receive thousands or tens of thousands of technical support calls per day, reducing calls by just one or two percentage points results in huge cost savings.

If you are considering a strategy based on call deflection, consider how a typical user attempts to solve a technical problem. Typically, they will:

1. Ask a friend

2. Use Google

3. Call technical support

You have two opportunities to interrupt this process; the first one is when the user asks a friend. If the friend has the relevant knowledge, the user gets the needed information. Mark Baker writes in "Why documentation analytics may mislead":[11]

> Much of the information that originates from your manual, then, may reach the user not through their direct reading of that manual, but via a network of mavens.

Baker goes on to explain that raw web site hits are less important than reaching the right people. In a group of 50 people, there might be one or two mavens. Mavens will disseminate information to the other 48 people. Thus, ensuring that the product mavens have the information that they need will potentially provide "ask a friend" support to the group that relies on those mavens.

There are a variety of ways to cultivate and support mavens. The company might provide:

- Early product access (beta testers)
- Premium technical support (mavens might be permitted to skip past level 1 support and go directly to more experienced agents)
- Discussion forums
- Recognition as a product expert (examples of this include the Microsoft MVP and Adobe Community Expert programs)
- Product updates with links to useful resources

The better the information that the mavens have, the more information they can provide to their network.

The second pillar of support is Internet search, especially Google search. Unless there is a compelling reason not to do so, technical content should be available on the Internet. Otherwise, when users search the Internet (and they will), your content will not show up. Examples of "compelling reasons" include the following:

- The information is classified (secret government information).

---

[11] Every Page is Page One blog (everypageispageone.com/2011/12/07/why-analytics-may-mislead/)

- The information is export-controlled or sensitive (information about encryption protocols or maintenance instructions for a nuclear power plant).

The argument that technical content contains proprietary competitive information must be balanced against the following issues:

- Competitors will find a way to get the information, perhaps by simply buying the product.

- Requiring user authentication to see the information greatly cuts down on the number of customers who will look at the information.

- What is the cost of not making the information available via search? How many people will *not* log in to the corporate site to find the information, and will instead contact technical support?

If you cannot bring yourself to publish the content where Google and other search engines will find it, you should still publish on a private web site and provide access credentials as appropriate to your customers. At that point, the responsibility for providing an excellent search experience shifts to your web site manager.

More efficient support operations

Technical content is not just for customers. It is also a valuable asset inside your organization, particularly for the technical support team. The technical support staff is responsible for helping customers to use a product successfully. The technical support organization desperately needs technical content—and the sooner it's available, the better.

Technical content that is locked down in PDF files and print is not particularly useful to support agents. Support agents need quick answers to questions, so they need the ability to search all technical content quickly. They might be able to use the index or table of contents of a single printed book to find what they want, but given a library of books, an online search is the only reasonable option. With PDF files, the process looks like this:

1. Search the company web site or intranet for a particular PDF manual.

2. Download and open the PDF file.

Loading...please wait

3. Search the PDF file to find specific information.

By this point, the customer has been on hold for several minutes and is probably steaming mad. The support agent isn't very happy, either, because she is expected to handle a certain number of calls per hour, and this lookup process is killing her numbers.

A better alternative is to provide content on a web site. Support agents can do a quick web search, display the information that they need, and help the customer. If the technical documentation group chooses not to provide an efficient way to access their information, the support staff is likely to look for workarounds, including the following (all these examples actually happened):

The support staff set up a wiki (a web site with many contributors).
On the wiki, they first copied and pasted parts of the official documents out of the PDF files. They made updates where the copied information was out of date or where they disagreed with it. Eventually, the technical support staff began writing additional content and posting it in the wiki. Meanwhile, the technical documentation group was producing carefully vetted information that was reviewed by subject matter experts. The information in the wiki was not reviewed and was often inaccurate, but the wiki search was usable, and that immediate availability overrode any other concerns.

The support staff created a knowledge base intended for technical notes.
Technical notes are supposed to be applicable to technical support but not needed in the product documentation; for example, notes on specific product configuration problems. Support staff then began writing basic procedural documentation and storing it in the knowledge base. The information in the knowledge base duplicated and sometimes contradicted the information in the official documentation.

The support agents created private content hoards on their local systems.

This eliminated the wait for a download, but the content caches had to be updated manually, which usually did not happen. The support agents could search faster, but were searching out-of-date information.

When the technical support staff sets up a parallel content authoring system because accessing the technical content is too much work, you have a serious problem. These shadow documentation efforts are a symptom of the larger information-access problem. To avoid them, it is critical to provide an efficient way for technical support staff to search and access information. Otherwise, you face the following operational problems:

- *Copying and pasting.* The process of copying and pasting content takes up time the support team should spend on much more important activities (such as keeping end users happy).

- *No easy path for updates.* When the official technical content is updated to reflect changes in a product, does that information show up in support information? And if updates do make it into the unofficial support information, it's likely copied and pasted, which is more wasted time.

- *Contradictory information.* With multiple copies of the same procedure, inconsistency is hard to avoid—and the result is confused customers and readers.

Technical content should provide a firm foundation for the technical support operation instead of undercutting it or making it more difficult. The role of technical support staff in creating content varies in different organizations, but duplicating content because the provided file formats are unusable is an appalling waste of resources.

Content collaboration across the organization

Historically, technical communication, marketing, and other departments developing content did not work closely with each other.

Each group offered the same justification: "Our content is completely different than theirs. We need to work on our own."

Today, collaboration is increasing. Instead of working in isolation, departments are identifying shared content assets and pooling their resources to develop this content more efficiently.

Technical communication and marketing usually "own" large amounts of content. But there are other, less obvious places where high-value information is created:

- *Product design and development.* Product specifications, marketing requirements documents, and other design documents exist before products are created and often include information needed for technical and marketing content. Product developers are usually involved in content development, either as contributors or as expert reviewers.

- *Training and education.* For instructor-led training and e-learning, instructional designers develop student guides, instructor guides, job aids, and more. These training materials often use step-by-step instructions and other technical content. In our experience, about 50 percent of instructional content is (or should be) identical to information found in tech comm.

- *Technical support.* Technical support staff is usually the heaviest internal user of technical content. As end users call in with problems and questions, the technical support staff must find the answer—and quickly. In addition to using technical content, the technical support team often creates condensed "cheat sheets," frequently asked question lists, troubleshooting procedures, or other information. Too often, this content is not contributed back to the technical communication, instructional design, and marketing teams.

- *Software.* Software products often contain technical content. For example, many complex software systems have extensive error messages. These messages should be documented and explained in the technical content. If the software and the error message explanations are generated from a single source, you can ensure that both sets of information are synchronized.

- *Online help.* Software engineers create unique identifiers for various interface components. The technical communication team creates content and matches the content up with the identifiers to enable context-sensitive online help. The online help files are developed by the tech comm team and then included in the software builds.

- *Product interface labels.* These are needed for hardware and software products. For example, a smartphone typically has a button (either physical or on a touchscreen) for making phone calls, labeled **Call** or **Send.** A common challenge for smartphone manufacturers is that this button has dozens or even hundreds of variations—at least one per supported language and often additional variations demanded by the cell phone carriers within a single language. And, of course, each smartphone has many of these types of labels. One solution is to store all of the labels in a database and extract the correct label based on the current language and carrier setting. This requires close collaboration between the engineering and technical content teams.

- *Web services.* The web services team manages the presentation, organization, and distribution of web content. Technical content published to the corporate web site should match the look and feel of other web site content.

- *Sales.* The sales team relies upon technical and marketing content for proposals and sales support materials. For example, if a potential customer has questions about specific features of a product, a salesperson may put together a custom package of information by compiling pertinent sections from user manuals, data sheets, and marketing materials.

Sharing content across departments improves the overall product quality by ensuring that customers receive consistent information and a unified message. At the same time, content sharing eliminates redundancy, which reduces the cost of content development. The challenge with shared content is cross-departmental collaboration. It is time-consuming and often difficult to establish strong working relationships across disparate teams.

Figure 8: Collaboration across disparate teams

"Collaboration" ≠ "content free-for-all"

It's true that the tech comm and marketing staff usually have above-average writing skills. But isolating content in a protected silo is not the answer. The "unwashed masses" may be less skilled than the writers at crafting euphonious[12] sentences, but they have useful information to contribute.

Collaboration doesn't necessarily mean that anyone with access to a keyboard should be able to create, modify, or release content. Content collaboration can be managed. For example, a collaboration system can:

---

[12] This extra-snooty word accurately reflects the attitude of many writers toward their non-writer colleagues.

- *Codify review cycles.* Send out notifications that request input on content within a specific timeframe. Track whether reviewers have offered input as requested.

- *Enable levels of access.* Determine whether a particular employee can review, edit, create, and distribute information.

- *Manage the approval and release of content.* Include mechanisms that indicate when content was approved, who approved it, and when it was released to end users (internal or external).

A functioning collaboration system allows the organization to manage contributions from writers with deep technical knowledge—even if their writing skills are relatively weak.

Chapter 3: Marketing and product visibility

One basic marketing tenet is that potential buyers need to be aware that your product exists before they can buy it. Technical content can contribute to—or detract from—marketing efforts.

Supporting marketing with technical content

Tech comm and marcom have long occupied opposite ends of the content spectrum. The stereotype is that tech comm is text-heavy, dense, and badly formatted whereas marcom is shiny, beautiful, and content-free. From there, the debate just intensifies:

Table 1: Marcom versus tech comm: the stereotypes

	Marcom	Tech comm
Design or automation?	Design	Automation
How much detail?	As little as possible	As much as possible
Assumed impact on revenue	A lot	None
Primary purpose	Persuade people to buy	Inform people
Writers are product experts?	No	Yes
Customer interaction with content is measured?	Yes	Rarely

Table 1: Marcom versus tech comm: the stereotypes (continued)

	Marcom	**Tech comm**
Affects product positioning?	Yes	Yes, but not on purpose
Important?	Yes	No
When do people read?	Before buying	After buying

The reality is not so simple. Some information products fall into a gray area between the two disciplines. White papers, for example, are full of technically detailed information but are intended to be persuasive.

Data sheets offer factual product specifications, but are generally used as part of the sales and marketing process. The purpose of a data sheet is to provide a potential buyer with specifications on a product so that the buyer can determine whether this product is appropriate for her needs. But a paper data sheet is not necessarily the best approach to this problem, especially if your product has a lot of possible configurations.

Instead, why not offer a web-based form that asks for some information, narrows the options down to a manageable number, and then lists those products in a chart for easy comparisons? Companies often have all of the information needed to do this, usually locked away in a printed or PDF data sheet.

Stash all the specifications in a database (in fact, the product design team probably has a database already). Then, create a web interface that allows the buyer to query the database to see which product makes the most sense for their requirements.

Enabling the buyer to "query the database," however, does not mean giving the user a literal database query experience:

Figure 9: Configuration tool that will not get used

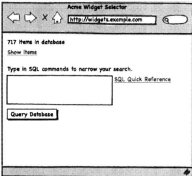

Instead, give them a friendly interface that lets them quickly narrow down their options and choose the one they want. You don't have to expose all of the fields in the database—just the ones that help people narrow down their choices.

Figure 10: A better database selector. Notice that the word "database" does not appear. The product list on the right updates as you make selections on the left.

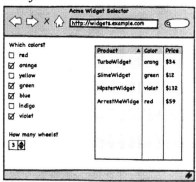

One powerful way of unlocking business value in your technical content is to rethink its presentation and its usage. In many cases, technical content has marketing applications, and by providing a more user-centered approach to the content, you can increase the value of the content.

Reinforcing the marketing message

You can use technical content to buttress your marketing claims. For instance, if you claim that your product is easy to use, back up that claim with a quick start guide. WordPress provides a good example of this with their repeated references to a five-minute installation process.

Figure 11: The "Famous 5 Minute Installation" for WordPress (codex.wordpress.org/Installing_WordPress, captured March 21, 2012)

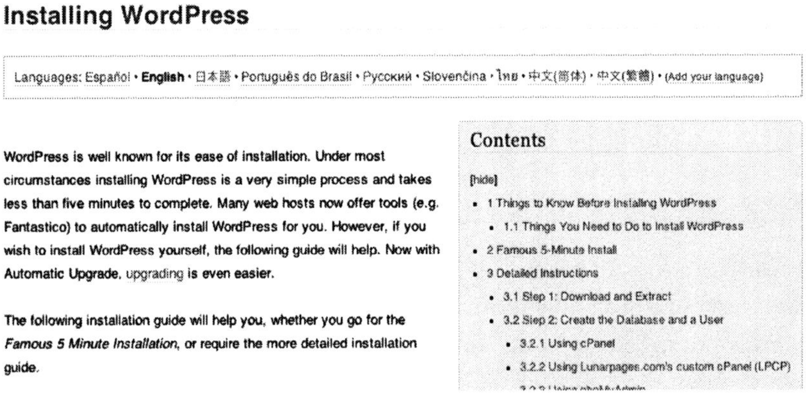

Installing WordPress

Languages: Español • **English** • 日本語 • Português do Brasil • Русский • Slovenčina • ไทย • 中文(简体) • 中文(繁體) • (Add your language)

WordPress is well known for its ease of installation. Under most circumstances installing WordPress is a very simple process and takes less than five minutes to complete. Many web hosts now offer tools (e.g. Fantastico) to automatically install WordPress for you. However, if you wish to install WordPress yourself, the following guide will help. Now with Automatic Upgrade, upgrading is even easier.

The following installation guide will help you, whether you go for the *Famous 5 Minute Installation*, or require the more detailed installation guide.

Contents

[hide]

- 1 Things to Know Before Installing WordPress
  - 1.1 Things You Need to Do to Install WordPress
- 2 Famous 5-Minute Install
- 3 Detailed Instructions
  - 3.1 Step 1: Download and Extract
  - 3.2 Step 2: Create the Database and a User
    - 3.2.1 Using cPanel
    - 3.2.2 Using Lunarpages.com's custom cPanel (LPCP)

Motivated buyers nearly always have the ability to review technical content, such as product documentation, before they buy. If your marketing claims are not supported by the technical content, you have a problem.

When technical content contradicts marketing

If your technical content contradicts your intended marketing message, you are in good company. Many organizations have this problem, often because of these factors:

- Marketing and technical communication groups are in different parts of the organization.

- Marcom and tech comm do not coordinate their efforts.

- Marcom and tech comm staff do not respect each others' work. (Marcom tends to view tech comm as a bunch of introverted nerds who can't communicate normally; tech comm thinks that marcom is full of liars who don't understand technology. Neither stereotype is *completely* incorrect.)

- Executive management does not believe that technical content is relevant to market positioning.

- Technical content creators cling to outdated but familiar ways of delivering content.

In this situation, you need to align the technical content with the marketing message by changing the message, changing the content, or both. Until this happens, your marketing efforts will be undermined.

Table 2: Aligning technical content with the marketing message

Marketing message	Misaligned content	Aligned content
Easy to use	User documentation is convoluted and full of obscure jargon	User documentation is easy to understand
Easy to install	Installation guide is 100 pages	Installation process and guide are short
Cutting edge	Technical content is delivered in a three-ring binder	Technical content is delivered in lots of different formats (text, audio, video, mobile, ebooks)
Personalized	Technical content is generic	Technical content is customized for different readers
Powerful	No advanced information in the technical content	Detailed technical content, customization scenarios, software API documentation
Fun	Content is ugly (or uses default templates), boring, and has no visual appeal	Creative delivery of technical content, such as comics, interesting visuals, entertaining examples, or edgy graphic design
Global audience	Technical content is available in only one language	Technical content is available in customer languages

Increasing product visibility

Technical content can help organizations increase the visibility of their products in the marketplace. Officially, technical content is intended for product customers—people who buy a product and then look at the documentation.

But one opinion poll indicates that about one-third of buyers[13] look at the documentation before buying a product, and the quality of the documentation will affect their purchasing decision.

Figure 12: An explicit call to review documentation before purchase (tipsandtricks-hq.com/wordpress-estore-plugin-complete-solution-to-sell-digital-products-from-your-wordpress-blog-securely-1059, captured March 19, 2012)

Affiliate Software Integration

Can be integrated with the **WordPress Affiliate** Software Plugin. So if you decide to boost your sales by introducing an Affiliate Program later then you just have to activate the other plugin.

Detailed Documentation

Scared of getting worthless product documentation after you purchase the product? Checkout the **shopping cart documentation** before you make a purchase.

Great Support

Tired of listening to fake support promises? Checkout **our customer only forum** to see how we handle product related issues (usually within 24 hours).

Other potential customers may not even be aware that your product exists, but are looking for key features that your product offers. If you make product content available on the Internet, you can pick up additional prospects. To reel in new prospects, your content must perform in three different ways:

- *Searchable.* Information must be available via an Internet search. That means putting the information online, and allowing Google and other search engines to index the content.

---

[13] "Consumer Feelings about Product Documentation," an opinion poll conducted online by Sharon Burton (sharonburton.com/wp-content/uploads/2012/07/ConsumerFeelingsBurton2012.pdf)

- *Findable.* Information must perform well for relevant keywords. That means paying attention to search engine optimization, delivering solid content, monitoring web analytics, and using keyword-based advertising.

- *Discoverable.* This term refers to information that has in-bound links, especially on social media. You can provide the initial links (for example, tweeting about content), but readers will choose whether or not they provide additional links (retweeting, posting on Facebook, or writing blog posts about your content). A reputation for providing excellent content increases the likelihood that people will link to your information and thus make it more discoverable.

A surprisingly large number of companies do not make any technical content available to the general public. Many are concerned about leaking proprietary information to their competitors. You must balance these concerns against the cost of withholding information. If you do not make your information available for Internet searches, you rule out any serendipitous discovery of your products.

Putting information behind a customer login or requiring email registration seems like a reasonable compromise. But David Meerman Scott estimates that "20 to 50 times more people download free content" when registration is not required. (This estimate is specifically for white papers.) [14]

You can assume that your competitors are motivated and will find a way to access your technical content, even if you have it locked away. Customers and prospective customers, however, are perhaps less motivated than competitors, so every obstacle between them and the content reduces the number of people who will eventually see the content.

Third-party books

Especially for niche products, product visibility is greatly enhanced by technical content written by an outsider. A third-party book authored by a high-profile industry expert makes the product more credible. You

---

[14] webinknow.com/2011/10/new-b2b-lead-generation-calculus.html, accessed March 29, 2012

could directly subsidize a book, especially if your market is currently too small to make a book financially viable for the author. If you don't want to get involved directly in a third-party project, many of the techniques described in the section on user community and loyalty can help nurture potential authors.

It is possible or even likely that an outsider will criticize your product. The best response is to engage in a discussion, rather than attempting to stifle their point of view. For example, tell an industry leader that a missing feature is under development for the next version, and ask him or her for feedback on how to implement it. Remember that the outsiders often have more credibility precisely because they are outsiders and not employed by the product developer.

Building user community and loyalty

Customers expect to participate in a conversation about your products, although the level of community participation that you can reasonably expect varies for different industries. But even in highly regulated, restricted environments, the user community can offer critical feedback. Your community's motivation level is important—you want to invite users to participate, but you do not want them to feel as though you are sloughing off content responsibilities onto them. Open source projects are built on community participation, but many of them also have terrible, outdated technical content precisely because only the community is providing content. (A notable exception to this is WordPress, which provides excellent technical documentation via the WordPress Codex. [15] Because WordPress is a publishing platform, we speculate many people in the community have good writing skills.)

The general purpose of a user community is to engage users—to shift people from passively consuming content to participating actively in evaluating or modifying content. In increasing order of engagement, they might:

- Read content

- Rate content with stars, likes/dislikes, or votes

---

[15] codex.wordpress.org/Main_Page

- Attend an online event, such as a webcast
- Send email with comments
- Write a public comment
- Edit or update content
- Write original content
- Attend a user conference

Remember that only a small percentage of users will participate. The general rule of thumb is that for every 100 users, 90 are lurkers (passive readers), 10 will participate, and only 1 participant will be very active. [16] That implies that you need thousands of users to have a successful community.

Assuming that your user base is a reasonable size, you need to weigh the risks of user participation. First, if you are in a regulated industry, such as medical devices or pharmaceuticals, your organization may be held accountable for user content. Unlike other industries, you cannot simply disclaim responsibility for user-generated content in the web site terms of service. This rules out uncontrolled submissions, such as unmoderated (and often raucous) user forums. You might, however, consider allowing industry partners to create content that is reviewed before being made available to your customers. For example, a medical device manufacturer could ask physicians to contribute articles about how they use a device. These articles would be reviewed by the manufacturer before publication on the corporate web site or perhaps in a newsletter distributed only to other physicians.

If your product is generally not dangerous to use (for example, most software and consumer electronics), then the risk of allowing user content is minimized. For these products, you need to carefully consider the risks that come from *not* allowing user participation:

- Do your customers expect a community? Does failure to provide a community have negative implications for your product positioning?

---

[16] Jakob Nielsen, Alertbox, October 9, 2006, useit.com/alertbox/participation_inequality.html, accessed March 27, 2012

- If you do not provide a community, will a third-party community spring up over which you have no control at all?

- Are your competitors providing user communities for their customers?

If you have a vibrant community, you have:

- Access to product testers.

- Access to end users. You can ask them questions, send out surveys, and use metrics from user discussions to direct product decisions.

- Access to people who write additional content for you. They will probably focus on edge cases, and that is acceptable. You need to produce the basic technical content.

Differentiating user content from official content

If you allow users to contribute content, be sure that user-contributed content is clearly differentiated from corporate content.

Figure 13: User-added text is flagged with a change bar and an author name

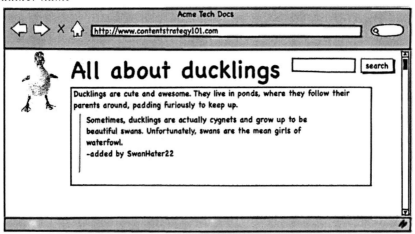

This approach helps you to maintain the distinction between official content and user content.

Extending the game experience into content

The computer game industry has been at the forefront of collaborative authoring and user-generated content. Many games come with minimal documentation that explains only how to operate basic controls. After that, it is up to the player to explore the game world. In online role-playing games (such as World of Warcraft), there are hundreds of locations to explore, objects to find or buy, and monsters to fight. Even without explicit support from the game developer, the user community often gets together to create reference information for the game.

For gamers, providing too much documentation can actually work against the gameplay experience. It might be wise, though, to provide support for the community to create their own documentation, perhaps by delivering a wiki with page stubs (starting points) along with user forums where players can discuss the game to their hearts' content.

Figure 14: Star Wars The Old Republic wiki example (swtor.wikia.com/wiki/Jedi_Knight, captured February 29, 2012)

Chapter 4: Legal and regulatory issues

Meeting legal and regulatory requirements may not be particularly cutting-edge, but the rationale is crystal-clear—failure to do so can result in product recalls, the loss of the ability to sell a product, and other "interesting" problems.

Avoiding legal exposure

The risk of legal liability for product documentation varies greatly across industries and locales. That said, products that can harm people are of greater concern than others. Farm equipment with sharp blades carries higher risks than a smartphone. A software application that controls a medical radiation–dispensing device carries higher risks than a computer game.

But as you explore how to handle legal requirements, what *really* matters is the opinion of your organization's legal counsel. If the legal team wants two pages of at the beginning of each document warning about the possibility of decapitation by manual can opener, they will probably get their way.

Unfortunately, most of the legal document priorities are directly at odds with creating useful content. An organization can focus on eliminating any legal risks in the documentation, and for a potentially dangerous product, this might even be appropriate. But if instructions are buried under extensive warnings, readers may instead turn to disclaimer-free (and maybe unreliable) third-party content. The risk is that maximizing legal protection may also minimize content consumption.

Note: None of the information in this section is a substitute for advice from an actual legal professional with expertise in your specific industry and jurisdictions. (See? We all need disclaimers.)

Intellectual property, patents, and limiting product claims

Intellectual property considerations may require content restrictions. Your legal team may ask you to limit the amount of information you provide, or require that readers must log into a corporate site (and agree to the site's terms of service) before viewing information. Establishing a registration requirement means that:

• Your web content will not be visible to Google and other search engines. Visitors cannot discover the information through a standard web search.

• The number of people who visit the site will be reduced.

If there is litigation regarding a specific product claim, the legal team may provide you with guidelines on how to document that claim.

Warnings, cautions, and other admonitions

Some industries use standardized warnings. In the United States, for example, the ANSI warnings labels are common. There are some generally accepted principles for admonishments—for example, Danger indicates a possibility of death or injury. For software products, these standards are sometimes modified to indicate the risk to data rather than humans.

It is common to see a raft of warnings at the beginning of a document. From a legal perspective, putting warnings at the front of a document makes the document more defensible. From a reader's perspective, it probably makes more sense to put the warning very close to the step that is potentially problematic.

Trademarks

Here are some examples of trademarks used incorrectly:

• I Skyped my brother last week.

- I googled, but I couldn't find anything useful.

Cease-and-desist letters notwithstanding, it is unlikely that you can stop people from misusing trademarks in this way in casual conversation or in social media. For more formal content, however, trademarks should be used correctly (even though it is highly awkward):

- I used Skype to call my brother. (Or, for the real sticklers: I used Skype software to call my brother.)

- My Google search didn't find anything useful.

Verbifying (or nominalizing) trademarks is frowned upon by the trademark owners.

Most technical documents have a list of trademarks at the beginning of the book, which looks something like the following:

> Oracle, JD Edwards, PeopleSoft, and Retek are registered trademarks of Oracle Corporation and/or its affiliates. Other names may be trademarks of their respective owners. [17]

A more detailed version from Microsoft:

> Information in this document is subject to change without notice.
>
> Companies, names, and data used in examples herein are fictitious unless otherwise noted. No part of this document may be reproduced or transmitted in any form or by any means, electronic or mechanical, for any purpose, without the express written permission of Microsoft Corporation.
>
> © 1995–96 Microsoft Corporation. All rights reserved.
>
> Microsoft, MS, MS-DOS, ActiveX, AutoSum, Bookshelf, Encarta, FoxPro, FrontPage, IntelliMouse, IntelliSense, MSN, Microsoft At Work, Microsoft Press, Multiplan, Outlook, PivotTable, PowerPoint, Rushmore, Visual Basic, Windows, Windows NT, Wingdings, XL and design (the Microsoft Excel logo) are either registered trademarks or trademarks of Microsoft Corporation in the United States and/or other countries.

---

[17] docs.oracle.com/cd/B19306_01/server.102/b14200/title.htm, captured May 30, 2012

Apple and TrueType are registered trademarks of Apple Computer, Inc.

The Mac OS Logo is a trademark of Apple Computer, Inc. used under license.

MathType is a trademark of Design Science, Inc.

Genigraphics and In Focus Systems are registered trademarks of In Focus Systems, Inc.

NetWare and Novell are registered trademarks of Novell, Inc.

Other product and company names mentioned herein may be the trademarks of their respective owners. [18]

As with all of these matters, you should consult your legal counsel to determine whether a blanket "other names may be trademarks" (as in the first example) is acceptable or whether an explicit list of trademarks is required (as in the second example). If you are required to provide a detailed list, automate the production of this list, as it is in no way a value-added technical communication project. A few techniques that we have used are:

- *Default copyright/trademark list.* All deliverables get the same list of copyrights and trademarks, so the list is set up once and then reused.

- *Automatically generated list from explicit markup.* Each trademark that occurs in text is marked as a trademark (or registered trademark). The trademark list is generated from the marked-up items. For example:

```
We use <trademark type="tm">MathType</trademark> to
create equations.
```

- Automatically generated list from reference list: We create a universal list of trademarks in a separate file. When it's time to generate output, the trademark list is generated by filtering the global list to include only the items that occur in that deliverable.

The last option requires the most setup and the least ongoing maintenance work. The middle option requires the author to do some

---

[18] biggs.org/office/gettingresults/OFCCPYRT.HTM, captured May 30, 2012

ongoing work. The first option is the simplest and most appealing if you can get your legal team to agree to it.

In addition to the trademark list in the front matter, you may also need to mark trademarks in the actual text. The traditional rule is that you must indicate a trademark on the first occurrence of that item. Unfortunately, that works better in books than in online content, where "first occurrence" a bit hard to define. Your choices for trademark indicators are as follows:

- Do not mark any trademarks in text.

- Mark the first occurrence in a book, chapter, or web page (preferably in an automated way; otherwise, your authors will be annoyed).

- Mark every occurrence of a trademarked term (guaranteed to annoy your readers and, if done manually, infuriate your authors).

Meeting regulatory requirements

Regulatory requirements can be tedious, but at least the business imperative is clear: without documentation that meets the requirements, the organization will not be allowed to sell its products (or, in some cases, operate at all). Other business goals may exist but are secondary.

To begin, you need to understand the actual requirement. Some requirements are technical. For example, the U.S. Food and Drug Administration requires that content of labeling (drug information) be submitted electronically using the Structured Product Labeling (SPL) standard.[19] SPL is an XML standard. Any content strategy that involves these regulated submissions must support the delivery of content encoded in SPL.

Other regulations, such as those in the European Union for product documentation, establish standards for what information needs to be included, as described by TCEurope in *Usable and safe operating manuals for consumer goods* (2004)[20]:

---

[19] fda.gov/ForIndustry/DataStandards/StructuredProductLabeling/default.htm

[20] docbox.etsi.org/STF/Archive/STF285_HF_MobileEservices/
STF285%20Work%20area/UG/Inputs%20to%20consider/
TCeurope_Securedoc_1_Usable%20and%20safe%20operating%20manuals
%20for%20consumer%20DKE-AKDFA_2004-01-01.pdf

> European Union legislation specifies that a technical product is only complete
> when accompanied by an operating manual. Delivery or sale of a product
> without an operating manual or with an inadequate manual breaks the law. In
> this case, users are entitled to assistance.

In the United States, only a few categories of product documentation are regulated, but concerns about legal liability may drive content decisions. One notable exception is the nuclear power plant licensing process. The U.S. Nuclear Regulatory Commission provides detailed requirements for the content of a licensing application.[21]

Another category of regulation is for component manufacturers, such as in the aerospace industry. If you make aircraft components, such as emergency exits or seats, your product documentation is subject to the requirements of your customer (such as Boeing or Airbus). If your customer informs you that you must deliver product documentation using S1000D (another XML standard) or a military standard, you have no choice but to comply. (Technically, these requirements are not "regulations," but in your role as a supplier, the distinction is at best semantic.)

Regulatory requirements are best seen as prerequisites—they affect your content strategy decisions long before you can start thinking about how to create effective, useful content.

Information delivery

To ensure that you meet regulatory requirements for information delivery, you need to know what content must be included in your information products. This varies greatly by industry, regulatory agency, and locale. (For example, the European Union may have different rules than the United States.)

Once you determine the content requirements, you need a content strategy that ensures that you always deliver the required content. For new drug applications, for example, you may be required to include detailed information about clinical trials in a specific format. That implies advance planning years before the application is submitted to an entity like the U.S. Food and Drug Administration.

---

[21] nrc.gov/reading-rm/doc-collections/fact-sheets/licensing-process-bg.html

From a content strategy point of view, you want to make sure that the appropriate information is being collected, and is available to be processed into the information product required by the regulatory body.

Although reuse is encouraged in most technical content, it may be problematic in regulatory submissions because you do not want to change a document that has already been submitted. Therefore, you may need to break the link between the original file and the reused content to ensure that any updates do not change the source document.

Some compliance requirements can affect content strategy. The most critical ones are:

- *Audit trails for content changes and reviews.* You may need a system that allows you to audit changes after the fact. Some organizations use cumbersome manual processes to create audit trails, but software is available that can automate this process.

- *Change tracking.* You may need the ability to deliver content updates with all changes marked so that regulators can easily identify differences from one version to the next.

- *Traceability.* You may need to be able to show the roots of each piece of content, so that you can justify its accuracy.

Technical standards

Technical requirements primarily affect information delivery. There are, for example, U.S. government agencies that require information to be delivered in a specific format, such as SPL (XML standard for drug information) or MIL-STD-40051 (a U.S. Department of Defense standard for technical manual preparation).

Noncompliance is probably not an option, so you have just a few choices:

- Set up an authoring and publishing environment that creates the information products as specified.

- Create information products in a noncompliant format and then convert them to the required format.

- Hire a third-party vendor to create compliant documents from whatever you are creating.

Figure 15: Options for technical compliance

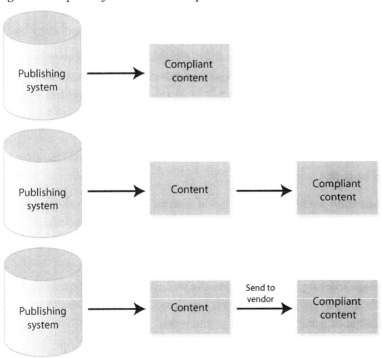

Many government regulations now specify XML as a required format. In the past, regulations were more likely to require a specific delivery mechanism, such as paper or PDF. It was left to the organization creating the information to determine how to create the final format. But today, the requirement for XML goes much deeper. Instead of focusing on document appearance (double-spaced, uses a specific font, and is printed on US Letter paper), requiring XML means that you must correctly encode the meaning of the information. For example, a part number might need to be presented in boldface (**22-44-66**). In the XML specification, there are at least two ways to tag a part number that result in boldface output:

```
<b>22-44-66</b>
<partno>22-44-66</partno>
```

Only the semantically correct approach (<partno>) would be acceptable.

Another example is Extensible Business Reporting Language (XBRL). The U.S. Securities and Exchange Commission requires reports in this

XML standard, which is used for financial and other business information. Here are a few lines, chosen at random from a Google filing (and somewhat simplified):

```
...
<FurnitureAndFixturesGross>65000000
</FurnitureAndFixturesGross>
<Goodwill>4903000000</Goodwill>
<IncomeTaxesReceivable>23000000</IncomeTaxesReceivable>
<IntangibleAssetsNetExcludingGoodwill>775000000
</IntangibleAssetsNetExcludingGoodwill>
...
```

Authoring in a standard office word processor with styles called Normal, Heading1, and so on will not result in information with this level of specific encoding.

Building a compliant publishing system

Setting up a publishing system that supports complex technical requirements is not for the faint of heart. This approach makes sense, however, in the following scenarios:

- A significant percentage of your business depends on compliant content. (For example, you are a defense contractor and are required to follow military standards.)

- You cannot afford any delays in content delivery. (For example, you are a pharmaceutical company, and you do not want a drug launch to be delayed because you are still converting over to the required format.)

- It is less expensive to set up a system in-house than it is to hire a vendor to do the work.

- Conversion is time-consuming and error-prone because the source files do not contain all of the information needed to generate the conforming files.

For authors, working in a specialized system will require training and support to help them understand the new workflow. Full-time authors are generally expected to master professional tools, but if your content is created by part-time authors who have other responsibilities, the use of specialized tools can be a challenge.

Converting to the required format

Instead of upending your established content creation process, you can add conversion as a caboose to the established publishing process. The main advantage to this approach is that content creators do not have to change how they work.

In some organizations, the content creators are subject matter experts (SMEs) with many other responsibilities. For example, a research scientist or a software engineer may be expected to contribute some content. For highly paid contributors with unique expertise, it can be sensible to accept content in a non-compliant format and leave the publishing tasks to support staff.

Most often, the non-compliant format is Microsoft Word. That is, content is created in Word files and must later be converted to another format. Some automation is possible, but you can expect that conversion will take a significant amount of time (several days or several weeks) after the document content is completed. In-house conversion makes sense in the following situations:

- The requirement for compliance is a one-time event. For example, your organization has a single government contract that requires delivery in a specific format.

- Content creators are unwilling or unable to change their workflow.

- Content creators' time is so valuable that a change in workflow is out of the question.

- In-house staff has the expertise to either build a conversion process or to do a brute-force conversion.

Other than the time required to convert the content, the biggest problem with a back-end conversion is that the source content may not include the needed information. Final documents may require metadata (such as the author, last modified date, and document type) that is not available in the source format. The only way to ensure that the information is available is to build a process that takes those requirements into account.

Offloading compliance onto a vendor

Some organizations go a step further by outsourcing conversion to the required format to an outside vendor. This is reasonable for a one-time custom requirement, but if you are creating a lot of regulated content over time, you probably want to keep the process in-house. Most vendors use a combination of scripting and brute-force retagging to get the content into the required format. Outside conversion is often expensive and time-consuming.

For government contracts, you may be limited to in-country vendors.

Part II
Developing a technical
content strategy

Chapter 5: Assessing the situation

With the exception of (very) early-stage startups, most companies already have content and content development processes. A thorough assessment of these information products is in order to ensure that you can avoid repeating any mistakes made in the past. It is also helpful to understand the gap between the strategy you need and the strategy currently in place. The burden of converting existing content over to a new system may affect your choices.

At a minimum, you need to examine the following:

- Existing information products

- Information development process

- Output requirements (old and new)

- Reuse opportunities

What you discover will help you determine the best way to support the identified business goals. The goal of this assessment is to develop a set of requirements for your content process. Those requirements help you determine which tools and technologies best address your situation.

> **Note:** An incremental approach is usually more feasible and less risky than discarding all existing content and starting over. However, if the existing content is spectacularly bad, starting over can be less expensive than attempting a rewrite. Signs that your organization might be a candidate for the scorched earth option include the following:

- Extremely outdated tools (WordPerfect, anyone?), which indicate that nobody has given the content development process any attention in at least a decade.

- Information products that do not have a clear purpose or audience, such as a user guide that alternates between beginner content and deep technical content.

- Incoherent instructions, such as procedures with dozens of steps or instructions embedded in paragraphs (rather than steps).

- Terrible writing with poor grammar, run-on sentences, and bad word choice, usually in a single sentence.

- Repetitive, mind-numbing details instead of useful technical information.

Analyzing existing information products

A content audit helps you determine what information you have and how useful it is. Begin with a list of current information products.

Table 3: Inventory of existing information products

Title	Pages	Last updated	Source format	Delivery format
Widget User Guide	400	2009	FrameMaker	PDF
Widget online help	400	2011	RoboHelp, but originally derived from User Guide	CHM
Widget Administrator's Guide	800	2012	Word	PDF
Widget API Reference	500	2011	Wiki (many pages are stubs without actual content)	Wiki
Video tutorials	20 minutes total	2011	Camtasia	Flash

A strategy for 2,000 pages is going to look very different than a strategy for 20,000 pages, so an order-of-magnitude estimate for page count is helpful. If you find yourself providing information about updates not by year but by month, that tells you that content is updated frequently. The update dates will also give you an idea of how much content is actively maintained as opposed to content that is static or rarely updated.

The source and delivery formats are interesting because they help you see what the status quo is. If 90 percent of your content is stored in a single authoring tool, your strategy looks different than if content is divided equally across four different (incompatible) tools. Often, the source formats reflect political divisions within the organization. The use of two competing, incompatible toolsets may be because of a merger (the technical content organizations never aligned their processes), a lack of collaboration (different managers chose different tools without consulting their peers), or simply different content requirements.

If your content is translated, you need a list of languages.

Table 4: Localization inventory example

Languages	What is translated?
French, Italian, German, Spanish	All technical content
Russian, Hebrew, Arabic	User guide and online help
Chinese (simplified and traditional), Japanese, Korean	API reference, user guides, and online help
(future) Greek, Turkish, Vietnamese	Planned for 2013: User guide and online help

In addition to the content produced by technical communicators, you may have high-value content produced elsewhere in the organization, such as the following:

- Training materials (instructor guides, student guides, e-learning content)
- Proposals
- White papers
- Knowledge base articles (usually technical support)
- Videos with tutorials

It is quite common for other departments to use technical content as a starting point, and you need to determine what that workflow looks like. A visual representation (even scribbled on a piece of paper) can help you identify problem areas where reuse occurs but is done inefficiently.

Figure 16: Technical content flow

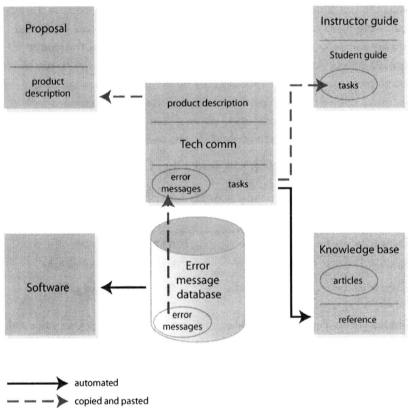

Those dotted copy-and-paste lines represent hours of work. You could further refine the analysis by looking at the volume, frequency, and work required for each arrow.

Table 5: Shared content assessment

Content	Quantity	Frequency	Destination	Method
Task information from Widget User Guide	200 pages	Major new versions	Student guide for intro training	Save FrameMaker to Word, then copy and paste
Error messages in software source code	600 error messages and explanations (message plus one paragraph)	Yearly	Admin guide	Copy and paste
Product descriptions	10 sentences	Weekly	Proposals	Sales team has their own copies; check for updates occasionally
Basic tasks	30 pages	Twice yearly	E-learning content	Save as XML; import

The preceding example would tell you that the task information for training and the error messages probably take a lot of time and should be high priorities for improved automation.

Your goal is to get a reasonable idea of the following:

- Total volume of content (page count, topic count)
- Types of content (books, manuals, videos, help, web content)
- Update requirements (yearly, quarterly, monthly, daily?)
- Versioning requirements (do some documents have variants, such as customer-specific versions?)
- Language support (what languages is content translated into?)
- Source formats (in what format is the content stored?)
- Output formats (in what format is the content delivered?)

You also need a more qualitative assessment of the content. You need to determine whether the content is:

- Accurate
- Current

- Complete

- Concise

- Audience-appropriate

After reviewing the existing information products, you should have a list of content challenges and ideas for improvement. Here are some common scenarios:

Reuse between technical documentation and training materials
The training department uses reference and task information created by the technical documentation team, but the instructional designers copy and paste instead of linking because the two groups use different, incompatible content creation tools. If the two teams standardized on a single workflow, they could share content seamlessly and avoid lots of tedious reworking.

HTML output is needed in addition to PDF
User guides and reference materials are delivered as PDF files with hundreds of pages. Readers, especially the technical support staff, would prefer HTML content because it provides a better search experience and faster access times.

Technical content is outdated because of inefficient updating process
The book production process—generating tables of contents and indexes, checking pagination, creating change page logs, and similar tasks—takes a significant amount of time. As a result, books are only updated twice a year. But the product is changing quarterly or even monthly, so the technical documentation is almost always out of date. Readers are complaining about the lack of synchronization between the product updates and the content updates.

Better content sharing is needed between technical content and technical support groups
The articles in the knowledge base and the information in the technical content often overlap or contradict each other. The groups creating that content need a system that enables sharing of information.

Content does not meet readers' needs
Content is poorly written and organized, not appropriate for the audience, or both.

Information flow

To determine how raw information becomes technical content, you need to understand and map the flow of information. Typically, you work backward from the final information product (perhaps a book or a web page) to determine the original data sources for text and graphics.

Figure 17: An example of information flow (typical for high-tech)

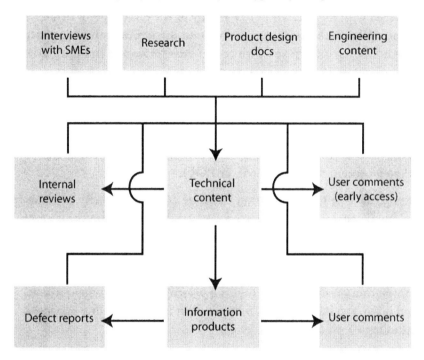

The information sources are the items at the top of the flowchart. You can further refine this information by specifying who is responsible for each step in the process and when responsibility is handed off from one group to another. For example, in some organizations, engineers write content drafts and give those drafts to the technical publications group so that they can "make it pretty." In other organizations, technical communicators write content independently with minimal input from the engineering team. Some organizations use a hybrid approach: perhaps the technical communicators write user documentation and the engineers write systems documentation. It is important to understand how information flow is currently working and where the problem areas lie.

Reviews are a common source of frustration. According to the authors, reviewers are lazy delinquents who give content only a cursory glance but try to prove that they did read the information by nitpicking the comma usage, usually incorrectly. According to the reviewers, authors tend to inundate them with 300-page documents, due back in two days, and delivered with no advance notice exactly at the busiest point in the release cycle. Both parties complain about inefficient, annoying processes for marking up documents with comments (reviewers) and integrating comments into the source content (authors).

The distinction between technical content and information products is not always present. In many web-based tools, for example, content is stored in its final form. In most publishing workflows, however, there is a transition between a source content format (maybe XML) and a delivery content format (HTML or PDF).

As you examine the information flow, the goal is to identify bottlenecks and inefficiencies so that a new process can improve upon the current workflow. For example, one common recommendation is to eliminate any content duplication via copy-and-paste and instead carefully manage reusable information. Speeding up the update process so that published information products are current is a priority for many organizations. And often, the integration of a localization workflow into a previously monolingual process is a challenge.

Other considerations may include existing tools, skills, and corporate culture. There's a big difference between Microsoft Word and wiki authoring. Staff who have lots of experience with print production may find a transition to automated document creation annoying—they like having the ability to tweak page breaks. If your corporate culture glorifies last-minute heroics rather than careful planning, you cannot implement a workflow that requires multiple formal signatures on reviews. (By contrast, if you are in a regulated industry, it is unlikely that you can avoid formal review and approval cycles.)

The analysis goal is to develop a solution that:

- Streamlines the flow of information throughout the organization

- Supports efficient content creation and delivery (which reduces costs)

- Maintains or improves the quality of the information delivered to customers

- Provides an authoring environment that is appropriate for content creators

Output requirements

In what formats do you need to deliver your information? Once upon a time, the only answer was print. Then for many organizations, the answer shifted to PDF. After that, it was PDF and HTML. Today, it's PDF, HTML, and mobile-friendly content. So, output requirements include both media (print, online) and deliverable formats (PDF, HTML, mobile apps).

Start with your existing information products and assess how well they support the business goals. For example, 600-page PDF files are often criticized for being hard to search and slow to open. Do you have customers (internal or external) who need a more lightweight format, such as HTML? On the other hand, if you are using PDF to distribute content to customers so that they can print exactly what you delivered, PDF may be the most appropriate option.

In addition to the actual delivery formats, it is helpful to look at the following factors:

- Do you need to deliver content variants, such as customer-specific information?

- Is there a requirement for a complex deployment, such as segmenting information to be delivered on your public web site and on a private web site only available to customers?

- What language should be displayed by default?

- What version of the content should be displayed by default?

- What sort of search functionality is required?

- Do customers need to see different information based on which product components they have licensed?

These issues affect your delivery strategy.

Armed with a list of delivery formats, you can refine your content delivery requirements. Typical requirements might look like the following:

- Installation instructions: print

- Getting started content: print with high production values and/or interactive tutorial

- Procedural instructions: PDF and web/HTML content, available to the general public

- System administrator information: PDF and web/HTML content, available to customers with login credentials

Table 6: Output requirements

Content	Print	PDF	HTML	Mobile	Other
Installation	x				
Getting started	x				Interactive tutorial
Procedural		x	x	x	EPUB, Kindle
System administration		x	x		
Top 10 issues				x	Video
Advanced techniques					Podcast; wiki
On-the-fly configuration				x	

In addition to the high-level requirements, you will need to create detailed specifications for each type of output. For example, what video formats will you deliver and where will the video be stored? Will you publish it on a YouTube channel? Do you need to deliver information that is tailored to a specific customer? Do you have a requirement to deliver content to a partner so that the partner can embed the information in their content and rebrand it?

The output requirements will drive the information architecture and then the content authoring, management, and publishing tool selection.

Content reuse

Reusing content lets you reduce content development costs while simultaneously improving the quality of the information. Everyone is familiar with basic reuse—copying and pasting from one document to another. But copying and pasting creates disconnected copies, which then must be updated separately (or recopied after an update). If an organization needs to reuse more than a small amount of content, a better strategy is needed.

Here are a few types of reuse:

- *Linked content,* where one copy of the information is referenced in multiple locations. For example, a basic topic, "What is a relational database?" might appear in the introductory materials for numerous information products created by a database company. Instead of copying and pasting, the information is stored in a single location, and all of the information products point to the same topic. Reused graphics are nearly always linked.

- *Common or boilerplate content,* such as copyright pages or warnings. A standard warning (WARNING: Choking hazard) must use identical wording throughout all information products, so it makes sense to provide it in a central location—and restrict changes.

- *Reuse across mixed content types,* where information is used in different formats (and even departments). For example, a mobile phone manufacturer might have a database of user interface strings (such as **Call** or **Send** or their translated equivalents) that are used to build mobile phone software menus. The technical content group could also access this database to insert UI strings into their content. When (not if) the UI designers change the interface labels, the technical content changes with them. Another example of mixed content is the creation of a database for software error messages, which are then delivered both inside the software and in a reference document.

- *Reuse across departments,* such as technical training and technical communication. The challenges in this area are tied less to technology and more to people. For instance, it is quite common for technical training groups to use technical content as a foundation for classroom training or e-learning materials. They insist, though, that the information needs to be rewritten to meet their requirements. A better strategy is to write content that both groups can use efficiently, but this requires close collaboration and mutual respect.

Note: Ann Rockley and Charles Cooper have an excellent, detailed discussion of reuse in their book, *Managing Enterprise Content: A Unified Content Strategy* (ISBN 978-0-321-81536-1). The discussion here owes much to the reuse taxonomy they have established.

Reusing information is a powerful way to reduce the overall cost of publishing customized content. Many content strategy projects are justified based on cost reductions via reuse. Effective reuse also drives down the cost of localization because it reduces the total amount of unique content.

Incumbent tools

The initial assessment should include an inventory of tools and technologies that are already in place. The incumbent tools have advantages over new tools—no new purchases are required and staff is already familiar with the tools.

However, the content strategy should be driven by the business goals and not the capabilities of the incumbent tools. Using existing tools can reduce the cost of training and of software licenses, but it also tends to result in incremental rather than radical changes. If your information products are currently effective, this may be acceptable. If not, a drastic change in workflow may be the needed shock to the organization.

Chapter 6: Architecting a solution

The solution you develop needs to address several related disciplines:

- *User experience*—how the content consumers interact with the information you provide

- *Graphic design*—the look and feel of the information

- *Information architecture*—the information structure, hierarchy, and metadata

The main question you must answer is, "How can content best support the business goals?" As part of that, you need to think about reader characteristics, such as their level of literacy, language proficiency, motivation, technological expertise, technology access, culture, location, and other demographics. From there, you can move on to think about effective uses of:

- Media (web, print, audio, video, and so on)

- Message

- Form

- Function

You can create a variety of information products to support the business goals. They include:

- Books

- Mobile apps

- Catalogs
- Online help
- Web sites
- Podcasts
- Video
- Simulations
- 3D renderings
- User forums and other user content

The question is, which combination of information products gives you the best results—achieving the business goals and providing value?

The traditional user manual, a printed tome with a huge table of contents, extensive index, and aggressively awful formatting, may convey that your product has a certain gravitas (also called the "thunk" factor for the noise the book makes when dropped), but it does a poor job of actually giving information to readers. Hardly anyone wants to slog through a 600-page reference book. On the other hand, delivering reference information in a searchable, linked, interactive format lets a reader quickly locate the information they need—assuming that the reader has access to a computer or mobile device. Readers aren't necessarily aware that there are 599 pages of other content—just that they found the information they needed.

But text-heavy web sites are not the only or even the best answer. For some content, a short video might make more sense. For other information, perhaps a quick reference guide on laminated paper is appropriate.

Delivering content online is assumed to be less expensive than creating printed books, but crafting a compelling online experience is difficult. Don't just throw content at the web team and assume that all will be well. And remember the problem of airplane help (can the reader get information while cut off from the Internet on an airplane?) and readers for whom Internet access may be restricted (retail clerks), unavailable (mining operations), subject to censorship (governments plus many corporations with web filtering), or excruciatingly slow. A strategy of providing help tips and tricks on Facebook works only if your readers can access Facebook while using your product or service, and are actually Facebook users.

The goals of the solution phase are to answer the following questions:

- Which content do users need?

- How will users get the needed content?

- What is the best workflow to develop and deliver the content?

Your strategy needs to evolve over time as users add new ways of accessing content, such as mobile devices, tablets, and ereaders. In addition to immediate requirements, then, your ideal workflow needs to be flexible and scalable so that you can change it later to support new requirements. Designing for today's problems may impair your ability to solve tomorrow's problems.

You can capture the design decisions in any way that makes sense to you. For instance, a company that makes business software might decide on the following:

- Deliver core user documentation on a dedicated web site (HTML) and provide an option for PDF download of content.

- Ensure that web content is mobile-friendly.

- Begin planning for a tablet app for offline access, to be delivered in 12–18 months.

- Personalize content based on user profile (if the user is logged in).

- For high-end customization, provide a wiki where users can develop and share examples. Build out preliminary structure for wiki.

- Localize the core user documentation, but not the wiki content.

The actual deliverables need design implementation, such as:

- Figuring out the look and feel for the web site and the PDF files

- Creating a search interface for the content

- Building a mobile experience

When the solution phase is complete, there needs to be solid understanding of the various content deliverable types and their presentation.

Content modeling

Content modeling is a critical portion of the implementation. In this phase, you identify your organization's requirements, develop a taxonomy (classification system) that meets those requirements, and consider where metadata should be allowed or required.

> **Note:** Do not overlook the importance of metadata. Metadata is critical to making your content manageable—with or without a content management system.

The content model specifies how information is organized. In unstructured authoring tools, such as FrameMaker and Word, the content model is usually specified in a template and a style guide and is enforced by technical editors. In structured authoring tools, the content model is enforced by the software.

As you define the content model, you must balance precision and simplicity. Defining with precision leads to large, complex content models. Keeping the content model as simple as possible makes it more usable. Other workflow components, especially content management systems or single-sourcing plans, may put limitations on how you define your content model and what metadata you create. For example, using HTML5 as your content model is easy (everything is predefined), but it has limitations (for example, no provisions for content reuse). A custom content model can match your requirements exactly, but it is a lot of work to build it. Fairly early in the project, you need to decide whether an existing content model (such as an XML standard) is an acceptable fit for your content or whether you need to create your own content model.

Content development for technical information is in the middle of a difficult transition from unstructured to structured content, generally based on the Extensible Markup Language (XML). XML offers significant advantages, such as the ability to enforce consistent structure and the ability to generate a variety of output formats automatically. However, establishing an XML-based content development environment is technically challenging and often expensive. Several XML standards are available that support technical content. The Darwin Information Typing Architecture (DITA) offers a framework for modular, topic-based content with heavy reuse, and is being adopted by many software companies. Some organizations build custom XML content models so that they can design a model to their exact specifications; others decide that DITA is good enough for their purposes.

Factors to consider in evaluating whether unstructured or structured content (and within that, DITA) is more cost effective for your organization include technical requirements, cultural fit, licensing costs, implementation effort, flexibility, and the size of the organization.

> **Note:** Scriptorium's unofficial rule of thumb is that an organization has a business case for structured content (based on localization cost savings) if they have:
>
> - Ten or more writers
> - Four or more languages
> - More than 2,000 pages of content per language

Be careful using existing content as a starting point for your content model. The advantage is that you can build out a content model that supports everything you are currently doing. The disadvantage to this approach is that what you are currently doing may not be what you should be doing. Therefore, you need to take a hard look at the existing content and consider how well it meets your requirements. Also, think about how requirements might change in the future.

At the end of the content modeling phase, you will have a detailed document that describes the proposed content model and explains your decision to use (or not to use) a standard. You may want to include a flowchart or hierarchical tree diagram that explains the structure. The delivery medium is less important than the ideas conveyed.

Figure 18: Excerpt of a structure analysis flowchart (simplified model of DITA structure)

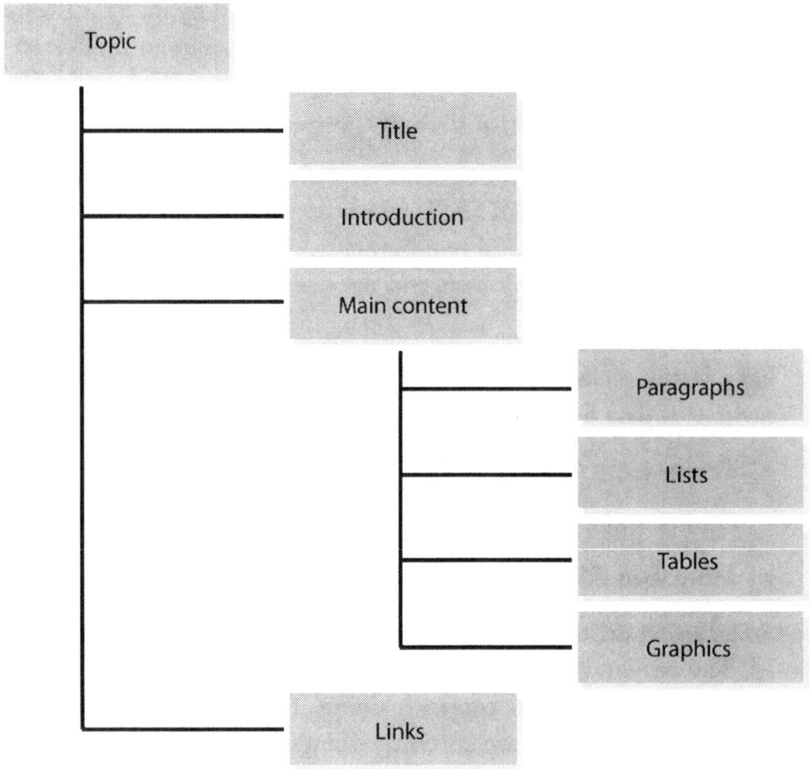

After review and approval of the content model, you can begin to build out that content model in the authoring and publishing environment. Remember that content modeling changes get more and more expensive as you get farther into the project.

Solution components

The organization needs to determine what tools and technologies it will use to produce the needed information products.

Content and digital asset management systems

A content management system (CMS) provides you with the ability to control text content with storage, versioning, support for reuse, and check-in/check-out features. Digital asset management systems provide additional specialized functionality for graphics.

The choice of content model will affect which CMSs you evaluate. CMS requirements include technical factors (such as hosted versus installed systems), usability, functionality, and price.

There are many options in this space. Some systems provide both the authoring tool and content management capabilities. Others provide only content management and must be integrated with separate authoring tools. Publishing tools are generally separate from the CMS and authoring tool but can be integrated with both.

Not every organization has enough content to justify a CMS. They may choose to use other options, such as:

- *Source control systems:* Although they are intended for managing software code, source control systems provide some useful file versioning capabilities for technical information.

- *Document management systems:* The distinction between a CMS and a DMS is that the content management system controls the components that make up a document (such as topics or reusable paragraphs). A DMS controls a document (such as a PDF file). Most technical content needs a system to manage components rather than a system to manage the final information products.

- *File system:* Smaller groups can use network drives and other rudimentary sharing mechanisms to ensure that authors can collaborate on files. These systems are always fragile and are not appropriate for organizations that are large, distributed, or subject to regulatory supervision.

Authoring tools

If a CMS is implemented, the choice of authoring tools is constrained by what the CMS supports. Without a CMS, you return back to the list of required information products to determine which authoring tools can meet your requirements. The top priority is to use an authoring tool that supports the required content model and output requirements.

Publishing tools

The publishing tools generate the needed output from the source content. In some cases, there is no distinction between the source format and the output format; for example, if you use a wiki, the wiki markup is rendered to the readers. An XML-based workflow, however, has XML as the source format and then a transformation process to create HTML or other output formats. Word processing and page layout tools create editable files, which are then saved or exported to PDF.

Chapter 7: Managing risk

Risks that are common in technical content include the following:

- Inaccurate information
- Incomplete information
- Information that is ignored
- Poorly written information
- Unclear information
- Information that is inappropriate for the target audience
- Legal liability due to documentation problems

Is there risk in having a really good content strategy? Consider the following factors:

- If information is easy to find and analyze, people may be able to more easily identify deficiencies in the products.

- Some companies like to practice security through obscurity. If the information is sufficiently difficult to locate, they reason that nobody will find it, and therefore the information is safe.

- Implementing a system that tracks and manages content changes may be seen as a litigation risk, because the record of changes might become discoverable.

Risks associated with the development of technical content include:

- *Release schedules.* Pay attention to product release schedules. Avoid scheduling launch of a new content development environment inside the chaos that is a release deadline.

- *Communication.* Good communication about the project alleviates fear, uncertainty, and doubt. Lack of communication does the opposite. The staff needs to understand the reasons for changing the approach to content and the benefits it provides. People do not like change. A careful change management process based upon a great deal of communication should address this challenge. When in doubt, overcommunicate—undercommunicating can kill your project.

- *Training.* Writers need training to understand the new workflow. Without training, they will take longer to learn the new process and may resent the steep learning curve.

- *Productivity.* Productivity will be initially low in the new system as people learn how to use it.

- *Quality of implementation.* The new workflow should closely match the requirements of the content that's being developed. Imposing processes that do not accommodate writers' legitimate requirements will lead to disgruntled writers.

- *Leadership.* Within the workgroup, the attitudes of leaders—whether positive or negative—will influence reactions of the entire staff. Without support from leaders, you will encounter heightened aversion to change and perhaps even outright hostility. Leaders may or may not be managers; look for the employees to whom others go for advice.

Most often, the risk perception is that implementing a new process is scary—and therefore bad. There's some truth to this. If you push through a new strategy and it fails, you may be risking your job. However, the alternative is that bad content will become a millstone that drags down whoever is responsible for it.

An effective content strategy will include an assessment of the risks along with a plan to mitigate them.

Corporate risk

Corporate culture affects what types of risk a company is willing to accept. A successful content strategy is compatible with the organization's risk culture. Long-time employees can often identify approaches that will not be successful, even though they may not be able to articulate the exact problem. To understand a company's risk culture, consider the following:

- *What is the company DNA? How long has the company been in business? How many offices? How many employees?* A company with a history of acquisitions operates differently than a company with a history of organic growth. A company that has 90 percent of its employees concentrated in the United States looks at the world differently than a company with 30 percent of operations in Asia, 30 percent in Europe, and 30 percent in the Americas. Some multinational companies are still distinctly a product of their home culture.

- *What does the company produce? What is the risk of using company products?* Financial services companies tend to be risk-averse and security-conscious in their content strategy. Game companies are often small and entrepreneurial—their priority is to be fun and cutting edge. Medical device makers are extremely concerned with product quality and health privacy issues. Heavy machinery producers worry about safe operation and usage of their products.

The risk priorities vary depending on the industry and the individual organization. Assess your organization to determine which risk factors are acceptable and which are not:

- *Cutting-edge technology.* Is the organization an early adopter or a laggard in implementing new technology? Where do the various systems being considered fall on this curve?

- *Open source software.* Is the organization dedicated to using open source software, completely opposed to it, or somewhere in the middle?

- *Cloud computing policies.* What is the organizational stance on cloud computing? If your organization has a policy of not allowing externally hosted content, that rules out cloud-based systems unless you can change the policy (which is risky and highly unlikely to succeed).

- *Audit trail.* How important is the ability to audit content changes? Is the organization in a regulated industry? (Note that this may vary by country.)

- *Change resistance.* Are employees and management generally willing to try out new systems, or are they change resistant? Have other projects failed because of change resistance?

- *Globalization.* How concerned is the organization about supporting global markets?

- *Quality.* How concerned is the organization about content quality?

- *Market positioning.* What is the product market positioning? Does the company sell low-cost products or premium products? Does the content strategy support the market positioning?

Technology risk

The risk of implementing new, unproven technology looms large in most content strategy projects. A bad technology decision can derail or destroy an otherwise compelling project. However, it is our experience that poor technology choices are rarely the cause of a failed project. Bad project management is a much more likely culprit.

Technology risks include the following:

- *Bad fit.* It is critically important to choose tools and technologies that solve your organization's problems. The fact that a particular system offers much better support for Feature X is interesting only if you actually need Feature X. A system that works miracles for another organization may not be appropriate for your organization.

- *Deficient products.* Some products do not deliver what they promise, or a feature is implemented in a way that does not solve the problem you need to solve. You can avoid this trap with pilot projects and other field testing. Pay attention to industry scuttlebutt—which vendors have good reputations? Which vendors are heavily criticized? Are the criticisms relevant to your requirements?

- *Customer support issues.* You think it is a missing feature. The vendor may insist it is a support or custom configuration problem. How does the vendor handle customer support? Are current customers generally pleased with the vendor? What happens when (not if) you uncover a bug in the product? Good customer support will mitigate the inevitable bumps in the road; bad customer support will magnify them.

- *Scalability.* Is the technology designed to handle the scope of content you need? What if you are wrong about the scope and you need to scale up significantly? For example, what if an unexpected acquisition triples the amount of content you are managing? Can the product handle additional content, languages, outputs, authors, reviewers, and so on? Will the cost of the product increase?

Rational requirements

As you define your content strategy, you can identify requirements for your technology layer. A good set of requirements will help you narrow your choice of tools to something manageable. Requirements should be measurable and be closely related to your strategy and therefore your business goals. For example, assume one of your goals is to control localization costs by reducing the amount of manual formatting done in production. You need a workflow that provides for efficient exchange with localization service providers, automates formatting, and supports all of the languages you need or expect to need.

Building requirements that tie back to your strategy will help you if you decide to change the toolset.

Tenacious ties to tools

Many content creators have *strong* opinions about the tools they use. Their ferocious (rabid?) loyalty presents an implementation challenge.

A high level of proficiency in a specific tool fosters a sense of achievement and security in team members. But a strong attachment to a tool can cause "tool myopia." Symptoms of this affliction include:

- Automatically rejecting other tools

- Using a current tool's capabilities as the benchmark for excellent technical content

- Defining new process requirements based on the current tool's capabilities

- Focusing on short-term productivity losses associated with new tools instead of considering the long-term gains

- Seeing new requirements through the lens of whether the current tool can accomplish them efficiently

The straightforward (but admittedly painful) cure for this myopia is building requirements. Strategic thinking about content cannot happen when early discussions are framed as *"Tool X can do this, and Tool Y can do that."* Focus first on the strategy; the tools come later.

For example, ask teams to evaluate what outputs (web pages, printed data sheets, PDF guides, online knowledge bases, and so on) content consumers need. Build wireframes and prototypes for these outputs, followed by more detailed specifications. How closely do the new outputs resemble existing outputs? The bigger the difference, the more likely that you may need new tools. Start the conversation about tools only after you thoroughly understand the requirements.

Include your current tools in the assessment to see how well they meet the requirements. It is possible that your existing toolset will support your new content requirements.

During consideration of process changes, content creators (including technical communicators, marketing staff, and instructional designers) often attempt to bias requirements analysis—usually, toward the incumbent toolset. Instead of an honest evaluation of the process of creating and distributing information, content creators jump on their favorite tools and start jockeying for position. By focusing on familiar tools, these employees make the creation and distribution of content about *themselves* and not what is best for the organization. A narrow focus on avoiding change makes the content creators less credible in their organization and diminishes the potential business value of content.

It can be difficult to get employees from multiple departments to focus on how technical content serves the entire organization, much less do that *and* steer conversations away from tools. Instead of managing these challenges on your own, consider hiring an experienced consultant to act as your requirements wrangler.[22] The consultant can:

- *Gather feedback from end users and departments that consume technical content.* External and internal content consumers are generally more honest when talking to a third party and not the content creators. Candid assessments of the current technical content are critical for developing requirements that maintain the positive qualities of existing content and improve upon its deficiencies.

- *Refocus conversations among content creators when discussions become tool-centric.* If content creators are displeased about the lack of tool talk, let the consultant bear the brunt of that unhappiness—better the consultant than you. A skilled consultant is accustomed to such reactions and should be able to diffuse any tension.

- *Offer guidance and suggestions on long-term goals for content.* As a manager, you may oversee the revamping of technical content just once in your professional lifetime. Consultants, however, often work on multiple content projects in a single year (sometimes concurrently). The expertise consultants gain from their past projects can be an invaluable asset to your efforts—and help you develop requirements for improvements you wouldn't have considered on your own.

- *Compile a report to codify requirements.* Designating the report as a deliverable from the consultant gives you a specific milestone for payment (which will make the accounting department happy), and it can provide you with a checklist for evaluating tools during the implementation phase.

  Note: To break up the costs associated with hiring a consultant, set up requirements development and implementation as separate projects. Many companies develop requirements in one quarter (or fiscal year) and begin implementation in another.

---

[22] We have no shame! We are consultants who recommend consulting services.

The continuity of having the same consultant work on both requirements and implementation is very beneficial *with the right consultant.* Having those two phases as separate projects offers you an escape hatch if your relationship with a consultant does not work out during requirements gathering.

Getting involved with your IT department

In the excitement of establishing a new content strategy, it's easy to overlook management and maintenance requirements for the required tools and infrastructure. Your information technology department, however, is painfully aware of these issues, and should be involved in the decision process very early on. IT can help you minimize (or eliminate) many technology-related risks.

Put aside the typical us-versus-them mentality. *"Really? Is the IT department going to lock that down, too?"* If you are going to put enterprise-level tools in place (as opposed to desktop applications), you are going to need help from IT. Early IT involvement is particularly critical if you are installing a content management system. To implement a new CMS, you will need information that goes way beyond the requirements of the teams who will use these new systems:

- How much time will it take to maintain the system (database maintenance, backups, and so on)?

- As more users and content are added to the system, how does maintenance time increase?

- Should the CMS reside on a physical or virtual server? How well will that server scale as users and content are added?

- Can the current Internet connection and network handle the CMS (particularly when multiple sites are involved)? Does the company need a bigger pipe to accommodate CMS activity?

- How will user accounts be added and removed from the system? Does the CMS integrate with the single sign-on solution your company already has in place?

- Will there be in-depth training provided to the IT personnel on the CMS? What about follow-on support?

- What is the process for installing new releases, and what kind of time do they generally require?

- Is there a hosted solution that eliminates maintenance tasks if the IT staff doesn't have the resources to manage another system?

Building a strong relationship with IT will be especially helpful when it is time to have The Talk. IT groups generally want to consolidate technology and to avoid proliferation of systems (both are very sensible goals), so they prefer to use existing technologies that could support your efforts. Sooner or later, they will ask, *"What about SharePoint?"*

If you have a detailed requirements document, you can work with IT to determine whether SharePoint meets the requirements. (Hint: Probably not. SharePoint is a document management system, not a component content management system.) Most technical content strategies rely on reuse at the paragraph and topic level, which is poorly supported by SharePoint.

Content creators and IT must work together to get the answers to critical technology questions—whether they like it or not! The content creators need to understand how their new tools affect IT, and the IT group must learn about how its efforts support specific content requirements.

Storage risk

Storage is really a component of technology risk, but is broken out separately here because of its importance. If you have ever had to migrate content from one storage format to another, with the attendant expense, difficulty, and problems, you are probably painfully aware of storage risk. Storage risk refers to the problem of file formats (and not to the actual file storage, such as hard drives or cloud servers). The format in which you choose to encode your content can increase or lower your overall project risk.

First, you must choose between proprietary and open storage formats. XML, for example, is an open standard, and valid XML files are (at least theoretically) portable from one XML editor to another. On the other hand, a proprietary binary format (such as older versions of InDesign or FrameMaker) is typically not portable to any other editing application.

The advantage of a proprietary format is generally a better editor; an open format provides more potential flexibility.

> **Note:** These days, the line between proprietary and open is blurring. InDesign, for example, provides InDesign Markup Language (IDML), an XML version of the InDesign binary format. IDML is complex because it includes all of the formatting and page geometry, but it is valid XML and can be processed with XML tools.

You can choose between formats that provide excellent semantic labeling and formats that are easy to understand. HTML, for example, generally falls in the latter category. Most flavors of XML provide better semantics, but are harder to read and understand than basic HTML.

You can store information in a database (and again, you can choose open or proprietary databases) or in flat text files. The former is more powerful; the latter is more portable.

The key questions you have to answer are:

- Where and how should we store information?

- How could we move information to a new storage system?

If you choose a tool with a high degree of lock-in, the success of your content strategy is yoked to the success of that vendor.

Chapter 8: Case studies

Content people, as a group, are terrible at business cases. In discussions about content challenges, the content creators usually have a litany of valid complaints about bad tools, inefficient processes, and generally wasted resources. But at the first mention of a potential solution that costs more than $100, content creators respond, "Oh, *they* will never approve that. We never get any money."

It is true that "they"—which is to say upper management—do not respond well to requests for large sums of money. What they do understand is business cases, such as, "To prevent us from wasting Y amount of resources every year, we need X amount of money this year," and X is smaller than Y.

Building a business case requires you to quantify how an investment (in tools, technology, training, or anything else) will improve business results. It is not sufficient to claim that your content strategy will contribute to business goals; you must estimate the improvement and show that the results are worth the investment. It helps when your estimates are obviously erring on the conservative side and the savings shown are still compelling.

Characters are composites—or entirely fictitious

This chapter offers sample business cases based on our consulting experiences. The scenarios, which use made-up or composite client profiles, describe common content strategy problems.

All of the client names are completely invented; any resemblance to an actual company or person name is unintentional. (We spent some time checking to make sure that these company names did not exist. We are not responsible for anyone deciding that Super Cygnets would be a fantastic name for a band.)

The examples show how to quantify potential savings and to compare them against the cost of implementing the needed system. For example, if you plan to increase the amount of reuse, you might need a content management system to support that effort. If you can show that a modest improvement in reuse in a large writing group will save $200,000 per year, you can easily justify a CMS-based system. If, however, your calculations show a savings of $15,000 per year, you need to scale your investment accordingly.

> **Note:** In many cases, you can show cost savings in multiple areas; a new content strategy can address multiple inefficiencies.

Estimating implementation costs

When estimating costs for a new system, examine factors such as the following:

- *Software and tools licensing.* Be sure to look beyond the initial licensing costs. Consider, for example:

  - The number of authors, contributors, reviewers, and so on, who will use the tool now and in the months and years to come

  - How much content will be managed now and in the future

  - The cost of software maintenance

  - The cost of upgrading to new versions

  - The costs to maintain and upgrade the hardware that runs installed tools

  - Annual/periodic costs for cloud-based solutions (software as a service)

- *Conversion.* If you need to convert legacy content to a new format, you can:

  - Do the conversion work internally

- Hire a consultant to handle it

- Implement a hybrid approach (for example, a consultant develops an automated conversion path, completes a pilot conversion, and then hands over the process so your employees can complete the majority of the conversion)

- *Training.* There is little sense in investing in new tools and processes if employees aren't shown how to use them correctly. There are several training options to consider, including:

  - Classroom-based training

  - Live (and recorded) web-based training

  - Self-paced training through workbooks

  - Computer-based interactive training

- *Consulting.* You can hire consultants to handle some or all aspects of a content strategy process—from requirements gathering all the way through conversion, implementation, and training.

- *Follow-on support.* If you need to make adjustments to your process after initial implementation, having a resource (often the consultant you hired to assist with the implementation) to help you with those changes is a good idea.

This list is not comprehensive; the factors you should consider will be based on your company's particular requirements.

The case studies offer general estimates on implementation costs. The scenarios illustrate the costs of typical projects but are not a substitute for your own organization-specific analysis.

Print workflow incompatible with new requirements

"Super Cygnets Corporation" has been producing printed manuals (yes, manuals) for their line of heat lamps since the 1960s, when people first started raising baby swans as a hobby. Somewhere in the 1990s, the organization shifted from printed manuals over to PDF files on a CD, which is included in the heat lamp box. But the entire content development process is geared toward producing sophisticated print/PDF deliverables.

The technical writers at Super Cygnets pride themselves on the high quality of the documents. They have complex graphics with embedded callouts, photographs, and more. They use drawings from the CAD system, but they modify them heavily to make them easier to read. Each chapter title page has a picture of a different happy swan family. Efficiency of content production has never been discussed.

The problem

Customers are demanding electronic deliverables. They want the ability to read documentation while they are standing in front of the heat lamp in their backyard swan coops. This means Super Cygnets needs to deliver information for tablets or smartphones. Super Cygnets is concerned about maintaining their position as the leader in baby swan heat lamps, which is a $5 million/year business.

The first attempt at a solution—pushing the existing PDF files onto a web site—is a disaster. The tables are very complex and impossible to render on a smaller screen. Graphics are high-resolution and do not render well on a smaller screen. The PDF files are huge, so they take forever to open, especially on a slower connection.

The Super Cygnets authors are experts at copyfitting and print production. They have no experience with other media, and they think that HTML is worthless because you can't control page formatting or your reader's fonts.

The solution

For information delivered on a smartphone or mobile device, you need adaptive HTML content, which will render nicely on a variety of devices. But the current workflow is intended to produce the best possible printed output and not HTML content. The source files are full of formatting overrides, so a conversion from the existing content directly to HTML will not yield good results.

Figure 19: Adaptive content

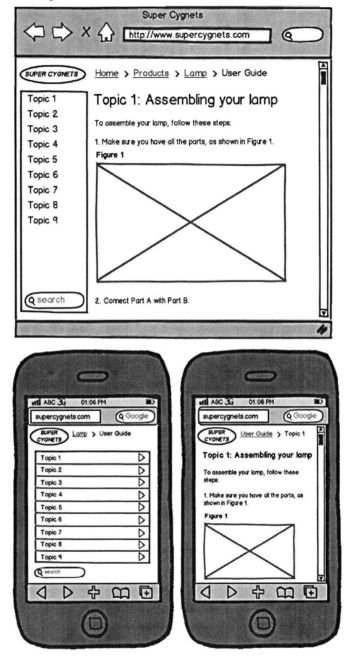

The answer is to do a few things:

- *Simplify the source content so that it will render properly in HTML.* Get rid of fussy formatting, and replace it with something attractive but straightforward.

- *Eliminate formatting overrides.* Educate authors on how those little tweaks that made the PDF files prettier are going to wreck their HTML output. (For example, the use of an invisible two-column table to present a long list of short bullet items in two columns.)

- *Evaluate the current authoring/publishing workflow and figure out where to introduce the option to generate HTML output.* Most page layout tools have a path into HTML; this may be a good interim solution that limits the disruption to the authors. Another option is to replace the page layout tool with something less print-oriented.

Figure 20: Publishing workflow for content on multiple devices

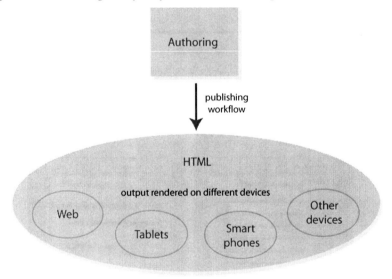

In a scenario like this, you can expect change resistance to be extremely high. Authors will resist a change in workflow that takes them out of their comfort zone (print production). They are immune to the argument that HTML is needed; they only see the negatives of diminished print quality. Some companies actually resort to eliminating the PDF output and making a clean break with the past.

The business case

Super Cygnets can continue using its current authoring software, paired with an HTML converter, so the software costs are smaller. The big expense will be the process of rewriting and reorganizing information to improve its implicit structure.

Table 7: Estimated costs

Item	Implementation cost (one-time)
Author training on writing modular content, using templates, avoiding overrides, and other single-sourcing concepts	$5,000
Rework content to eliminate formatting overrides, implement new template-based approach, and topics (1,000 pages x 30 minutes per page @ $50/hour)	$25,000
Clean up formatting template in print production tool	$2,000
License HTML conversion tool	$1,000
Design adaptive HTML output	$5,000
Implement HTML design in conversion tool	$5,000
TOTAL	**$43,000**

Super Cygnets cannot directly justify this expense with reduced costs, although there will be some cost savings from the use of templates rather than one-off formatting and page-by-page copyfitting. Instead, they are calculating that an investment in mobile and tablet-based content will increase their market share.

Table 8: Cost savings and revenue generated

Item	Cost savings and revenue generated (per year)
Less formatting time, two full-time writers (150 hours per writer per year @ $50/hour)	$15,000
Improved market share/competitive advantage of providing mobile/tablet documentation (0.5% of $5M product)	$25,000
TOTAL	**$40,000**

A more drastic solution, in which the entire workflow is replaced, would be very expensive to implement, incur enormous conversion costs, and be impossible to justify without assuming a steep market share increase (approximately 5%). One key to successful content strategy projects is to scale the solution to match the business drivers.

Executing an open source strategy

"Fledgling Research Inc." builds cloud-based applications based on open source technologies. Fledgling Research has some default building blocks in a code library, but most of the applications are custom-built and therefore the technical content also needs to be customized. And once the application and documentation are delivered, the customers often extend and build further customizations without the help of Fledgling Research. On average, each application is licensed for $10,000.

The problem

Until now, Fledgling Research's technical content has been in the form of sparse code comments. But as the company matures, the customers are getting larger and more demanding. These bigger customers are unamused by the lack of formal technical content.

The solution

Fledgling Research needs a content strategy that is compatible with the company's emphasis on open source tools. Fledgling Research decides that all content should be encoded in XML using one of the popular standards for technical communication (DocBook or DITA). All processing is done with open-sourcing tools, such as Ant, Java, and XSLT. Fledgling Research considers an open source content management system, but eventually decides to keep the technical content close to the software code instead. Technical content is stored parallel to the software code in an (open source) source control system.

The total software licensing cost? Zero.

Fledgling Research does, however, invest a significant amount of time and energy into building stylesheets that output the information in an attractive, professional format. The company then contributes these stylesheets back to the open source community so that others can benefit from them.

Figure 21: Open-source content strategy

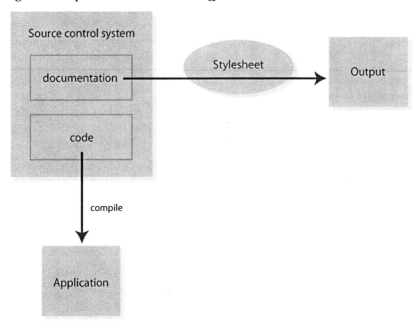

The business case

Fledgling Research has very few costs, and all work is done with in-house resources.

Table 9: Estimated costs

Item	Implementation cost (one-time)
Design attractive output	$5,000
Implement design	$15,000
TOTAL	**$20,000**

The cost savings and revenue sides of the business case are also sparse. One option is to assume that the increased customer satisfaction from better content will lead to at least one additional customer per year and that, likewise, the open source contribution will yield one additional customer per year. Another way to look at the business case is to assume that better technical content will open up projects with larger customers, which would increase the average value of a customer.

Table 10: Cost savings and revenue generated

Item	Cost savings and revenue generated (per year)
Increased customer satisfaction due to more professional technical content (one additional customer per year)	$10,000
Goodwill and increased visibility in open source community because of stylesheet contributions (one additional customer per year)	$10,000
TOTAL	**$20,000**

Coexisting with a suboptimal system

At "Vulture Chicks LLC," a leading venture capital and research firm, both technical content and marketing content are posted to the public-facing web site. The web services group has installed a content management system dedicated to web content, and they want both marketing and technical content to be stored and managed in the web CMS. The marketing group is working inside the web CMS, and it is a good fit for their requirements.

The problem

The web CMS does not provide features that the technical content team needs. Most critically, technical content is reused across various documents in ways that the web CMS does not support. In addition to web site content, the technical content group also creates output in PDF. Currently, the technical content group is authoring in a different system and then copying and pasting to move the information into the web CMS.[23] The process of copying and pasting information is taking up a huge percentage of the technical content team's time.

---

[23] I know we're going to be accused of inventing ridiculous, melodramatic scenarios here. My imagination isn't that good. This…stuff really happens.

The technical content group is being "encouraged" to move their authoring efforts into the web CMS. This would allow marketing and technical content to coexist in a single system, thus reducing the number of systems that need to be maintained. But the technical content group is resisting the change because they do not believe that the system addresses their requirements.

The solution

The best solution is to set up a content management system that meets both groups' requirements equally, but it is too late for that option. The new web CMS was expensive, and displacing it now is going to be politically impossible. Instead, the technical content group must coexist with this system. There are several options:

- Work within the web CMS, and adjust the workflow to what the system will support. That would mean eliminating PDF as a supported output format and eliminating most content reuse. The PDF requirement is negotiable, but the lack of support for component-based reuse (topics and paragraphs) is a showstopper.

- Work within the web CMS, but customize it to provide the features that the technical content group needs.

- Set up a separate system for the technical content group, and configure it to format and publish information that matches the web CMS output. This has the advantage of providing a system that meets the group's requirements, and the disadvantage of requiring two separate systems.

Figure 22: Separate systems for technical and web content

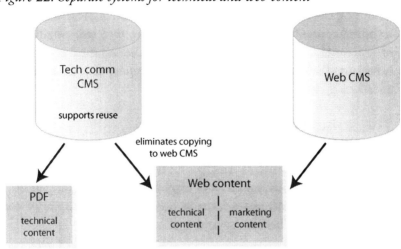

None of these solutions is particularly appealing.

After detailed analysis and discussions with a committee of Vulture Chicks executives, the following facts emerge:

- Customizing the incumbent web CMS to support component-based reuse would cost approximately $150,000.

- Customizing the web CMS to support better PDF output would cost approximately $30,000.

- The 10 staff members in the technical content group are currently spending 20 percent of their time moving information between incompatible systems.

- Of the 10,000 pages of technical content, 25 percent is currently reused.

Based on this information, Vulture Chicks makes the decision to leave the technical information in the current (separate) system rather than forcing the technical content group into the web CMS.

The business case

Maintaining redundant systems is never a preferred option. In this case, however, the marketing group chose a content management system that meets their requirements (strong templates and branding, web analytics, the ability to schedule content publication to coincide with product launch dates) without any consideration for technical content requirements (support for component-based reuse, PDF output in addition to HTML, strong versioning support).

Table 11: Comparing unpleasant options

	Web CMS	Dedicated technical content system
Customization for component-based reuse	$150,000	(already supported)
Customization for PDF	$30,000	(already supported)
Set up formatting to match web CMS	(already supported)	$20,000
Cost to maintain second system (10 hours per month @ $100/hour)		$12,000 per year

The component-based reuse is the biggest problem here, so it is worth asking how much value it adds.

Table 12: Cost savings and revenue generated

Item	Cost savings and revenue generated (per year)
Time savings from reuse, 2500 pages (30 minutes per page @ $50/hour)	$62,500
Cost of copying and pasting (20 percent of 10 staff members' time, 4000 hours per year)	$200,000
TOTAL	**$262,500**

In short, the copying and pasting process is an enormous cost and must be eliminated.

Table 13: Comparing business cases

	Web CMS with reuse/PDF	Web CMS without reuse/ with PDF	Dedicated technical content system
Customization for component-based reuse	$150,000		
Customization for PDF	$30,000	$30,000	

Table 13: Comparing business cases (continued)

	Web CMS with reuse/PDF	Web CMS without reuse/ with PDF	Dedicated technical content system
Set up formatting to match web CMS output			$20,000
Loss of reuse		$62,500	
Cost to maintain second system (10 hours per month @ $100/hour)			$12,000 per year
TOTAL	**$180,000**	**$92,500**	**$32,000**

Based on these numbers, a dedicated technical content system looks like the best of a bad set of options. One hidden cost is that the marketing and technical content groups will not be able to share information easily across systems. At Vulture Chicks, the potential reuse across those groups was minimal, but at a different company, that factor might change the business case calculations.

Supporting a component-based product

"Energetic Eaglets LLC" makes a suite of social media analytics tools. When the company got started, they had only one product—a tool for tracking business mentions on MySpace. But today, Energetic Eaglets offers a variety of products for dozens of social media tools, including blogs, Facebook, Twitter, and, most recently, Pinterest. Larger companies generally license the enterprise tools, but a web-based light version is also available for small businesses.

The problem

Technical authors have been using HTML editors to create separate sets of web pages for each product. However, that approach is no longer sustainable because the product variations have increased substantially. Currently, there are approximately 16,000 pages of HTML documentation. The 20 technical authors cannot easily determine if another author has already written content about a feature shared across product lines, so there are multiple versions of what should be identical content.

With customers licensing lots of different product combinations, Energetic Eaglets needs the ability to quickly deliver tailored web content to customers who can switch product licenses whenever they want—and the technical information users receive should change accordingly.

The solution

A component-based product line requires a component-based content strategy. For Energetic Eaglets, this means rethinking information as belonging to various products, identifying areas where there is reuse or overlap, rewriting information to maximize the reuse possibilities, and then developing a system that delivers information based on a customer's current list of products.

Figure 23: Component-based content strategy

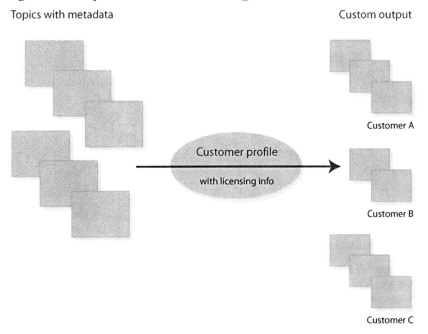

The solution requires the following:

- Component-based content that matches the architecture of the product lines.

- The ability to assign metadata to content that reflects product variants.

- The ability to filter information based on customer characteristics and product variants.

- The ability to generate customized information collections for each customer profile.

To meet these requirements, Energetic Eaglets will convert (and in some cases, rewrite) its content into XML-based chunks containing the metadata for filtering. The company will use a cloud content management system to manage the source files. Based on a customer's licensing profile, an automated process will dynamically create web pages from the source content.

The business case

Energetic Eaglets will implement a cloud-based content management system that has annual licensing, so cost estimates include both the one-time initial implementation costs and CMS licensing for the first two years.

Table 14: Implementation cost

Item	Year 1	Year 2
Cloud CMS (licensing for one year; includes web-based authoring tool)	$30,000	$30,000
Analyze content to determine metadata approach and shared information	$5,000	
Convert HTML source pages into component storage (XML) with automatic addition of metadata based on file names	$25,000	
Add metadata to XML topics where automatic conversion fails (2,000 topics x 5 minutes per topic @ $50/hour)	$8,000	
Design and implement dynamic HTML output	$25,000	

Table 14: Implementation cost (continued)

Item	Year 1	Year 2
Follow-on support (refinement of dynamic HTML output)		$5,000
Web services support for dynamic output (maintenance and tuning, 10 hours/month @ $100/hour)	$12,000	$12,000
Author training on component-based content	$6,000	
TOTAL	**$111,000**	**$47,000**

On the positive side, Energetic Eaglets realizes cost savings from reuse and the ability to deliver the required component-based content.

Table 15: Return on investment

Item	Cost savings and revenue generated
Elimination of maintenance of redundant content (2,000 topics @ 1 hour per topic @ $50/hour)	$100,000
Eliminate manual production of variants based on licensing changes (100 changes per year, 4 hours per variant, $50/hour)	$20,000
Add ability to change documentation immediately based on user profile and metadata instead of waiting for manual update, thus reducing technical support calls (2 calls fewer per change at $50/call)	$10,000
Improved customer satisfaction and market share based on sophisticated content support system that matches company positioning (for a $10M product, a 0.5% increase in market share)	$5,000
TOTAL	**$135,000**

Upgrading inefficient review processes

"Orbiting Owlets, Inc." manufactures a line of night vision goggles. The company distributes content in PDF and HTML formats on the Orbiting Owlets web site. During content development, technical authors send PDF drafts of guides to engineers via email to get feedback.

The problem

The technical authors at Orbiting Owlets send entire manuals in PDF format to the engineers and subject matter experts (SMEs) and ask for comments on specific pages. The engineers do not like receiving the large email attachments—or having to wade through hundreds of pages to find the particular content they should review. An engineer might receive a 200-page PDF file, in which only 20 percent is relevant for the review.

To send their feedback to authors, engineers use multiple techniques, including:

- Marking up hard copy
- Marking up the PDF file with the commenting tools in Acrobat Reader
- Compiling comments in a Word file
- Sending multiple emails as they read through the content

It is difficult for the authors to keep track of the input they received and how they modified content based on the engineers' feedback.

The solution

Orbiting Owlets needs an easier way to collect and track reviews. The company decides to implement an online review system that integrates with the authoring tools and a component content management system. This system enables engineers and other reviewers to add a layer of comments and changes to the source content. Technical writers then approve the changes to add them to the content.

Figure 24: Collaborative review in action

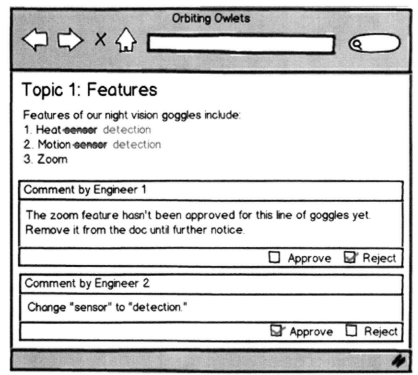

The new system does the following:

1. An author labels content for review by a specific engineer with a review due date.

2. The system sends the engineer a notification, which contains links to the content requiring review.

3. The engineer reviews the content. The system tracks the changes requested by the engineer.

   Note: If multiple engineers review the same content, they can see each other's changes.

4. The author reviews the changes, approves or rejects them (and offers reasons for rejections), and makes additional revisions based on the feedback.

Figure 25: Streamlined, collaborative review process

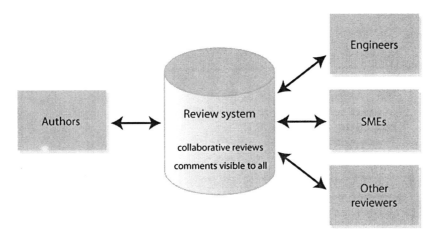

The review system sends reminders to individuals who have not responded to requests for feedback. The system can also compile reports about the review process: when authors requested input, when the reviewers offered feedback, and so on.

The business case

The justification for a new review system is based partly on review efficiency and partly on quality issues—better reviews should lead to improved content.

> **Note:** Orbiting Owlets is already using a source format that is compatible with component content management. This greatly reduces the cost of implementing the review system.

Table 16: Implementation cost

Item	Year 1	Year 2
Cloud CMS (licensing for one year; includes collaborative review)	$50,000	$50,000
Web-based training (2 hours) for reviewers to learn new system (30 reviewers)	$1,000	
TOTAL	**$51,000**	**$50,000**

Table 17: Return on investment

Item	Cost savings and revenue generated (per year)
Eliminate review PDF generation (5 PDF files per month, 1 hour per PDF, @ $50/hour)	$3,000
For engineers: Time saved during review by avoiding email/Acrobat markup, and more focused reviews (receive 40 pages instead of 200 pages; 20 pages per hour @ $50/hour; 8 hours saved per review)	$24,000
For authors: Time saved with integrated comment approval instead of copying information from PDF or email over to source files (15 minutes saved per change; average number of changes per review is 20; 5 hours of review changes per review @ $50/hour)	$15,000
TOTAL	**$42,000**

In this case, the cost savings do not pay for the system investment. Orbiting Owlets could either look for additional return (such as quality improvements because of better review cycles, faster time to market), scale down the content management system, or wait until increasing content volume justifies the system investment. As Orbiting Owlets is growing quickly, the company decides to make the investment in the review system on the assumption that increased content volume (and therefore review cycles) will justify the system in year two.

Backchannel content development

"SquabSafe, Inc." develops security systems that enable remote monitoring of homes. The technical communication group at SquabSafe writes content that becomes PDF guides and online help.

The problem

The PDF guides and help files contain some content that is useful to both end users and the technical support team, but overall, the official SquabSafe content does not include realistic examples or configuration scenarios that customers request when calling support. Also, the PDF and help formats are hard to search, which makes it difficult for the support team to find information during calls.

To make it easier to find information, the support team copies and pastes content from the guides and help into their own online support system. Support staffers then augment that information with real-life examples based on the calls they receive. They have also begun to write new content, often duplicating (and contradicting) information in the official documentation.

There is no mechanism to move content created by technical support into the official product content.

The solution

The root cause of this problem is that the formal content development process at SquabSafe is producing information that is not useful to the content consumers (in this case, the technical support staff). The technical communication group is hemmed in by a time-consuming review process and a lack of knowledge about the real-world examples that technical support needs. The technical support group ignores the official content because it doesn't meet their needs; the technical communication group resents the fact that the technical support group is producing rogue content without any quality control.

This situation, in which two (or more) teams duplicate information in order to address different requirements, is quite common. Correcting it requires much better alignment between the teams:

- *Eliminate duplicate content creation.* Writing the same content twice is a waste of resources. The technical support group will take on responsibility for examples and configuration scenarios, and will store them in their article database. The technical communication group will write core product information, and will incorporate changes and updates provided by the technical support team.

- *Improve findability of technical content.* The official technical content needs to provide excellent information retrieval options (search, indexing, findability, links, and the like). Content consumers (including the technical support staff) need the ability to filter according to product lines, product versions, support staff access levels, and so on.

- *Provide an efficient review and commenting system.* Because the support staffers spend a great deal of time talking to end users, they develop a deeper understanding of how customers are *really* using the product line. The support team will have the ability to review content under development and the official published information will have a commenting system, so that readers can provide feedback.

Figure 26: Solution to align tech comm and tech support teams

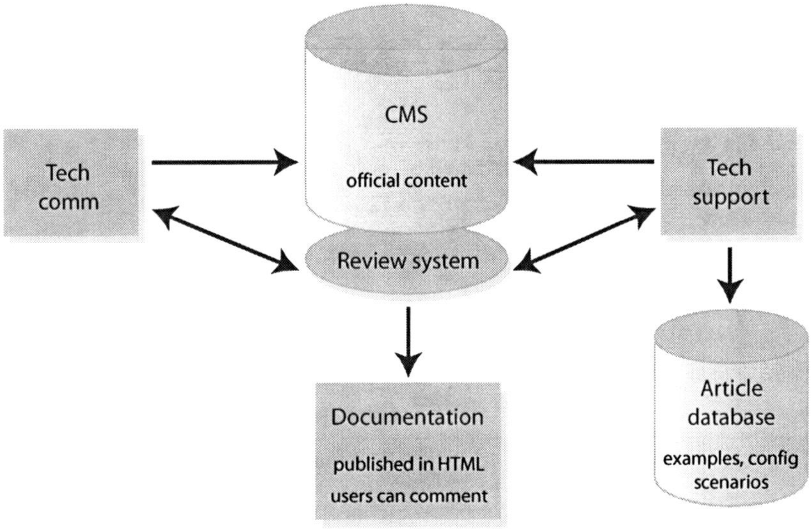

The business case

Internal customers, especially call center staff, spend a lot of time looking for information in technical content, which they need to pass on to customers. Improving the search experience and efficiency will result in cost savings in the call center. Assume that you have the following situation:

- Number of call center employees: 50

- Time spent for a single search in the current system (searching for content, opening PDF files, and searching again): 10 minutes

- Time spent in the new system (web search with HTML results): 30 seconds

Table 18: Cost estimate for creating HTML version of technical content

Item	Cost
New authoring/publishing system that supports single-sourcing and collaborative review	$100,000
Web site design, including desktop and mobile	$15,000
Set up automated process to create HTML from source files	$15,000
Configure search	$10,000
Training on new tools and technologies	$8,000
Content conversion	$100,000
TOTAL	**$248,000**

The calculations look something like this:

Table 19: Improving search efficiency for call center

	Current system	**New system**
Time required per search	10 minutes	0.5 minutes
Number of searches per employee, per day	10	10
Total time required per employee, per day	100 minutes	5 minutes
Time saved, per employee, per day		95 minutes
Time saved for all call center employees, per day		4,750 minutes (79 hours)
Cost savings per day (assumes $25 per hour for call center employee)		$1,975
Cost savings per year (200 working days)		$395,000

Even without calculating the time regained by avoiding content duplication, SquabSafe can easily justify the new system.

Improving time-to-market around the world

"Eyas Information, Inc." provides business intelligence services to customers around the world. Eyas is, however, growing quickly, and demand for its services (whatever they might be) is global. As a result, Eyas needs to deliver technical content in multiple languages.

Eyas has tried to limit localization costs by sending content to its in-country sales teams and letting them do the translation work. This approach looks reasonable from a budgetary point of view, but the result has been lengthy delays in translations, and the eventual results are of low quality. The company is now looking into using a professional localization services provider.

In this scenario, a strong business case will help to mitigate the sting of localization invoices. Currently, translated content (when available) lags behind the initial English text by at least three months. We can quantify the lost revenue of those three months, along with the potential customers lost because Eyas does not provide information in their preferred language.

The problem

Eyas has hesitated to enter new international markets because of the potential expense of localization. But a competitor, "Hawkeye's House of Hints," is beginning an aggressive expansion into worldwide markets. Eyas needs to ramp up localization to compete, or cede the non-English markets to Hawkeye. The rival companies are led by two men with long-standing animosity. Not competing is out of the question both for strategic and personal reasons.

The solution

Ultimately, Eyas wants to reach a point where new content is available in all supported languages at the same time. (The localization industry calls this "sim-ship" for simultaneous shipment.)

To begin delivering professional translations, add new languages, and (eventually) achieve sim-ship while keeping localization costs reasonable, the following needs to happen:

- Set up an efficient authoring process for source language content.

- Use best practices for creating localization-friendly content (simple sentence structure, no jargon, culturally neutral content).

- Ensure that formatting is completely automated (preferred) or heavily templatized.

- Use a localization services provider rather than ad hoc internal resources. (A professional in-house localization group is also a viable option, but only for organizations with heavy localization requirements.)

- Use topic-based authoring (rather than book-based authoring). Ship topics for localization as they are ready instead of waiting for deliverables to be completed.

Figure 27: Publishing workflow optimized for sim-ship

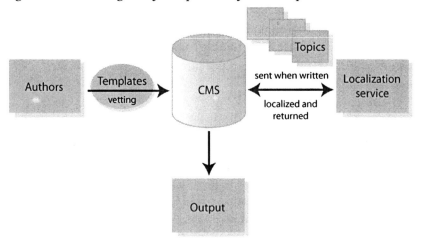

The business case

Table 20: Estimated costs

Item	Year 1	Year 2
Author training on writing modular content, using templates, avoiding overrides, and other single-sourcing concepts	$5,000	
Edit content to use best practices for localization (5,000 pages @ 10 pages/hour @ $50/hour)	$25,000	
Clean up formatting template in print production tool	$2,000	
Content localization (5,000 pages, 4 languages, $0.25 per word per language, average 250 words per page); 20% updates the second year	$1,250,000	$250,000
TOTAL	**$1,282,000**	**$250,000**

Clearly, the costs outside of localization will be a rounding error. After scraping themselves off the floor, Eyas management take a look at the gains from the more professional localization approach.

Table 21: Cost savings and revenue generated

Item	Cost savings and revenue generated (Year 1)	Cost savings and revenue generated (Year 2)
Faster time to market in 4 locations (delay reduced from 12 weeks to 2 weeks; assumes product revenue of $5M per year total across 4 locations)	$10,000	
Revenue growth due to providing professional content in the local language (10% of $5M per year)	$500,000	$550,000
TOTAL	**$510,000**	**$550,000**

With a professional localization process, Eyas takes the delay in localization from 12 weeks down to 2 weeks. That means that revenue starts flowing in 10 weeks earlier than before. This is worth approximately $10,000 (1% of $1M). Eyas sees a positive return sometime in Year 3 of this business case.

Reusing content from oddball sources

"Aggressive Gosling" produces a variety of highly technical documents. One document is derived from information in a product development database. The technical publication group at Aggressive Gosling periodically generates a report from the database, and updates the document based on any changes.

The problem

Generating the report is easy, but reviewing it for changes is dreadful and extremely time-consuming. Aggressive Gosling needs an automated way to get the information from the database into the related document. But the database is a homegrown project, and the output is custom markup that is unique to Aggressive Gosling. Aggressive Gosling needs a way to extract information from the database and publish it into a variety of formats using the established publishing workflows.

The solution

Before the rise of XML, the answer would have been to write a custom database-to-publishing tool connector. This answer is still viable, but it is usually more expensive than the XML-based alternative. So, the solution becomes to:

- Extract information from the database in whatever format is easiest

- Convert the extracted information into XML

- Integrate the XML into the existing workflow and use the established publishing tools

Figure 28: Using XML as middleware

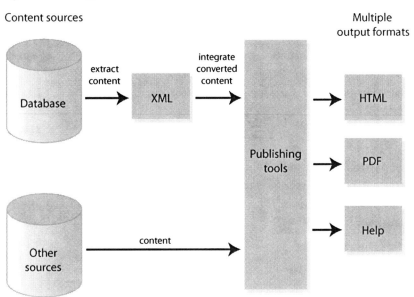

The business case

This is a straightforward case of eliminating tedious manual work with an automated, repeatable process.

Table 22: Estimated costs

Item	Implementation cost (one-time)
Write database extractor	$5,000
Convert database extract into XML	$10,000
Build automated import from XML into current authoring tool	$10,000
TOTAL	**$25,000**

Table 23: Cost savings and revenue generated

Item	Cost savings and revenue generated (per year)
Quarterly review of database extract to identify updates (10 hours @ $50/hour, 4 times per year)	$2,000
Manual insertion of updates into authoring tool (60 hours @$50/hour, 4 times per year)	$12,000
TOTAL	**$14,000**

The payback period here is just under two years, but it is also worth noting that a process that takes nearly two weeks (70 hours) per quarter is now done in minutes. The value of that increase in velocity is not quantified here.

Part III
Implementing your content strategy

Chapter 9: Methodology

Content strategy projects are no different from other large projects that involve process change. They are stressful, uncomfortable, and often difficult to move forward.

To implement your project and improve your chances of success, we recommend following these steps:

1. Identify and interview stakeholders.
2. Establish implementation goals and metrics.
3. Define roles and responsibilities.
4. Establish timeline and milestones.
5. Build the content creation system.
6. Convert legacy content.
7. Deliver content.
8. Capture project knowledge.
9. Ensure long-term success.

The sections that follow describe each of these steps.

Identifying and interviewing stakeholders

To ensure that your content strategy succeeds in the organization, you need to build momentum and support for the project. One proven way to do this is to identify the people who have the most to gain (or lose) from the new process. These are your stakeholders. You must involve the stakeholders early on in the project so that they have an opportunity to understand what is going to happen, and to ensure that their requirements are addressed in the content strategy plan.

At a minimum, your stakeholders must include the following:

Table 24: Content stakeholders

Functional area	Role
Content creators (authors, graphic designers, instructional designers, technical communicators, copywriters, and more)	Work in the new system to create content
Product management	Provide information about requirements for content products
Engineering	Provide support for integrating content into software or hardware; may review content
Information technology (IT)	Provide resources to manage and maintain the software systems
Information security	Sign off on security of chosen system
Executive management	Approve strategic direction and business case; decide whether to provide funding
Legal	Review legal implications, which may include the scope of audit trails, ability to roll back content to earlier versions, and how to ensure that a new system meets existing contractual obligations
Localization	Provide information about requirements for efficient localization; coordinate with localization vendors to ensure that new content strategy will be supported during translation

During the initial content strategy analysis, you should meet with these and other stakeholders to ensure that their input shapes the final recommendation. It is critical to identify legitimate constraints ("Our software runs on XYZ, so we must deliver in a specific format") and separate them from obstacles ("I don't like Windows"). Remember that the difference between an obstacle and a constraint may be the seniority of the person raising the issue.

Establishing implementation goals and metrics

The first implementation step is to identify your success criteria. Based on your business case, you can define what will make the project successful.

Table 25: Setting goals

Business case	Example success criteria
Localization cost	Hold localization cost steady while adding two more languages. Reduce lag time in localization content from four months to two weeks.
Improved search efficiency	Improve call center response times. Increase call volume by 10 percent without increasing staff.
Marketing support	Increase web site engagement.
User community and loyalty	Increase number of registered users and active users by 20 percent over the next year.
Increased collaboration	Add 15 active users from engineering community.
Content reuse	Increase percentage of reused content in source documents from 10 percent to 20 percent.

Each organization has different goals, expectations, and metrics. The key is to spell out the targets before the project begins.

The initial project specification should also include the following:

- A list of required deliverables and deliverable paths (for example, HTML created using XSL transformation on XML files)

- Any tool-specific requirements that affect the elements and attributes to be defined

- A high-level description of the planned workflow

After defining goals, you can develop success criteria. This allows you to evaluate the project's success when it is completed. Your criteria should include specific metrics. These might include items such as the following:

- Number of deliverables per writer before and after implementation
- Percentage of information reuse achieved
- Time required to do "final polish" on deliverables

Once the business goals are established, and you have defined measurable success criteria, it's time to consider who will do the actual implementation work.

Defining roles and responsibilities

The scope and complexity of the project will determine how many people are involved. At a minimum, you need to make sure that the following roles are covered:

- Education
- Development
- Review
- Approval

The education role is responsible for getting buy-in from all affected parties—especially managers who approve the effort and staff who will use the new system. Depending on the audience, you may use different approaches, such as presentations, informal discussions, and training.

If consultants are involved, they will most likely do the development work and then present it for your review and approval. There may also be an internal review on the consultant's side before you see any deliverables.

For larger projects, there is often a development team. For example, one person might be responsible for establishing taxonomy (element and metadata definitions), another for choosing and installing a content management system, and a third for creating output transformation stylesheets.

Our consulting practice uses a collaborative approach. We strive to identify or develop technical expertise on the client side as early as possible so that our clients can provide meaningful reviews and feedback as we build their systems. By combining our clients' business requirements and expertise in their own subject matter with our consultants' understanding of content strategy and publishing technologies, we can deliver a final product that improves on what either of us could do on our own.

Working on implementation and review teams will require significant time commitments from the participants. Any realistic resource plan must take into account other commitments, deadlines, and deliverables that could conflict with project requirements.

Establishing timelines and milestones

After defining the project's goals and resources, you can put together a timeline and milestones. These are tied to business requirements.

As in any project, establishing a schedule creates accountability. Without a schedule, the project may suffer repeated delays.

If you are working with a consultant, project milestones will likely be linked to incremental payments. Consultants usually do significant up-front analysis to ensure that they understand your project requirements before negotiating a contract. Implementing a new content strategy is expensive; even a medium-sized implementation can easily reach six figures with custom-developed documentation and hands-on training. Determining project scope before the project begins ensures that there are no unpleasant surprises for our clients later.

The timeline needs to include some slack for delays. The most common reason we encounter for lagging schedules is lack of reviewer availability, which causes delays in client reviews.

Building the content creation system

Rely on your implementation goals as a guide when you evaluate and select the tools for creating content.

During the evaluation process, it is tempting to primarily focus on the content creators' day-to-day work experience instead of the overall process goals. Successful implementation of the new system, however, depends on close attention to strategy and requirements.[24]

---

[24] From Scriptorium's *The State of Structured Authoring* (second edition), scriptorium.com/books/the-state-of-structured-authoring-second-edition

During your tool selection process, don't be dazzled to the point of distraction by vendor claims. Ask for evaluation licenses so you can thoroughly vet and test the tool. Talk to other companies using the tool, and read posts on product forums. You can also enlist the services of an independent consultant to help you pick the tool that best supports your requirements.

The details of implementing your system depend on the toolset you choose. If you select a document processing tool, you purchase the licenses, install the tool (or log in to a cloud-based application), and then set up templates for the process. For a bigger system that involves components such as a content management system, you must also configure the connectors (to a localization system, for example), establish the workflow within the system, configure the review process, and so on.

> **Note:** Early coordination with procurement and IT departments is crucial for smooth(er) implementation of larger systems.

Converting legacy content

A new content strategy may result in changes across several facets of content:

- File storage format
- Organization
- Writing style
- Output formats

It is usually possible to automate the conversion from an old file format to a new file format and the delivery of new and different output formats. But reorganizing and rewriting content requires the dedicated attention of a content creator and cannot be automated.

Most of our customers do at least some legacy content conversion, even if they are planning significant changes to their content approach. Here are just a few of the options:

- Convert everything into the new system.

- Identify high-priority content and convert it. For example, you might convert only content for the flagship product, or only for products that you expect will have significant updates.

- Just-in-time conversion. As new projects are scheduled, find the related content and convert it.

- Assess for conversion. Somebody reads legacy content page by page to determine what information is good enough to convert.

- Convert nothing. If updates are required to old content, use the old system.

It will come as no surprise that we recommend assessing the costs and benefits of these various options to determine your legacy content conversion strategy.

Figure 29: Just-in-time conversion strategy

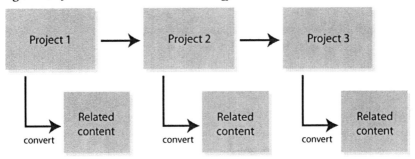

Content conversion is an ongoing, difficult challenge for many organizations. There are companies that specialize in this area with a wide variety of strategies. Some companies write custom scripts to automate as much as possible; others throw warm bodies at the project until they reach the required throughput to meet your deadline. Commercial and open source tools are available that support file format conversion.

Whatever strategy you choose, be aware that no conversion is perfect, and you will always need to ensure that there is a strong quality assurance/proofreading phase toward the end of the process. Some common conversion challenges include the following:

- Conversion is based on formatting that is present in the source documents. Documents whose formatting does not conform to the standard (formatting "exceptions") cause problems in conversion. Consistent use of templates/stylesheets in legacy documents makes conversion faster and more accurate.

- In some cases, the new content requires information that simply is not present in the source documents. For example, the new metadata structure always requires the author's name, but the author's name is not present anywhere in the original source file. Another problem can occur when you implement information typing—the various section types, such as Procedure, Reference, and so on, correspond to a single Heading2 tag in the original files. Assigning the correct information types may require manual intervention.

The best-case scenario for automated legacy document conversion is approximately 95 percent accuracy. Significant human effort is required to address the remaining five percent.

> **Caution:** We recommend against assigning document conversion to your staff as their introduction to the new content strategy.

Delivering content

The idea that content development and delivery can be separated is relatively new. Desktop publishing software combines development and delivery—page layout software creates a file that is ready for print delivery. In XML-based and multichannel workflows, however, the final content file does not necessarily resemble the final deliverable. For example:

- An XML file with approved content can be transformed into PDF, HTML, EPUB, and other formats. These various deliverables contain the same information, but they use different technologies and presentation.

- HTML content may include some presentational elements, but applying multiple CSS files to an HTML file can result in very different presentations.

- A responsive web site lets developers create HTML and CSS that is rendered differently on different devices.

- Even InDesign, which started out as a page design tool, has the ability to render EPUB, which could look quite different from the information presented in the InDesign authoring environment.

The delivery process, then, is the process of transforming the source files into the final output files. This could be a conversion from InDesign to EPUB, or a transformation from XML to HTML, or any number of other workflows.

Delivery is a key automation opportunity. Output that is lovingly handcrafted[25] will have nicer formatting than content that is generated automatically. The differences, however, are usually subtle and lost on the vast majority of content consumers. For technical content in particular, formatting needs to be usable. But the niceties of copyfitting, pristine kerning, and hand-tweaked hyphenation are unlikely to add enough business value to justify the cost. (Unless, of course, the target audience for the content is professional typographers.)

In this phase of implementation, you need to configure the system that extracts content from the content storage system and turns it into a deliverable format, such as paper, PDF, HTML, CHM, or EPUB. In some environments, this separation of authoring and publishing does not occur. For example, if you author in a web design tool, you might create HTML directly. However, in most modern workflows, you develop and store your content in a different format from the actual delivery format.

Once you have rendered the content into the final format (such as HTML), you must also make it available to your customer. In the print world, this meant printing and distributing books. In the digital world, this includes tasks such as pushing content from a test server to a live web site, pressing the Publish button in WordPress, and submitting information to the Apple Store.

---

[25] I *so* wish I had coined this phrase, but it belongs to Roger Hart.

Capturing project knowledge

The planning and documentation phases are thankless tasks. If done well, nobody really notices them. But if you skip them or give them less attention than they deserve, you will end up with a disorganized project that is impossible to maintain. We recommend that the project documentation should, at a minimum, include the following information:

- Content model explanations and recommended best practices for authors

- An explanation of all of the important project components, such as the content management system (which one? why chosen? what customizations?), the planned strategy for legacy document conversion, the authoring tools, known limitations, and so on

- In-depth technical documentation for developers, which will be used to maintain the system

- Formatting specifications and algorithms

It is also helpful to provide details about the project, background information, and the rationale for various design decisions.

Knowledge transfer is critical to a smooth implementation. Authors need training on the following topics:

- General authoring concepts (such as writing modular content) and using the designated authoring tool

- Project rationale

- Content model

- Metadata and taxonomy

- Best practices for authoring

We have found that authors accustomed to an authoring environment with established rules (templates, style guides, and the like) have an easier time making a transition to a new system than less organized groups.

If the staff that built the system will maintain it, little or no developer-level training is necessary. However, if the staff responsible for maintenance is new to the project, extensive training is required. In addition to the information needed by the authors, maintenance staff also needs the following:

- In-depth understanding of the content model, design decisions, limitations, and trade-offs

- How to implement and test changes

- Technical training on the output paths and how to maintain them

Ensuring long-term success

In an ideal world, you could build a new authoring environment, deploy it, and wash your hands of the project. In the real world, however, changes are inevitable. Even in the best-planned, most-organized environment, you will be required to make small changes to the content model, add new output paths, and so on. You must have a plan to manage these changes, as in any software development project. This means developing a change control process and identifying and prioritizing bugs and enhancements.

Your process must address several competing requirements: it must minimize changes, ensure that changes are made in an organized manner, and be flexible enough to ensure that the environment meets the workgroup's evolving requirements.

Some workgroups implement changes on a schedule. For example, for the first two years, the implementation team rolls out new versions quarterly; after that, every six months. You might also schedule change implementation based on priority—changes with higher priorities are done quickly; lower-priority changes are rolled into a scheduled update.

After building, testing, and deploying the project, you need to shift resources from development to maintenance.

As you move forward, you can evaluate the implementation against the goals you identified at the beginning of the project. Are you seeing increased productivity and reduced production-editing requirements? Is content management improving? If your implementation was successful, you can tick off the items you listed at the beginning of the project (and perhaps a few you didn't anticipate) as accomplishments now.

Chapter 10: Managing change

Handling process change is a significant hurdle for any manager. Good management is critical when a company changes workflow; without it, the implementation of new processes will likely fail. Bad management kills implementations, and things can get ugly for everyone involved.

So, what can a good manager do to ensure a smooth(er) transition when implementing new processes for developing and distributing content? At a minimum, take the following actions:

- Demonstrate value to upper management and those in the trenches

- Offer training and knowledge transfer

- Differentiate between legitimate issues with the new workflow and reflexive recalcitrance

- Enlist participants in a pilot project to explain process change

Demonstrating value

Showing cost savings is the primary way to get upper management to approve new content workflows (for example, demonstrating a significant reduction in localization expenses). Details about proposed processes and tools are generally a lot less compelling to those higher up on the food chain. When you're trying to get buy-in from upper management, money—specifically, spending less of it—talks.

However, talk of cost reduction doesn't warm the hearts of those who develop content. This is particularly true when authors are asked to give up processes they have used for years.

While acknowledging the accomplishments in the old process, an effective manager will show team members how the new process eliminates unpleasant, labor-intensive aspects of the existing process. Explain how the new process enables authors to focus more on developing good content instead of on secondary tasks (such as formatting). For most content creators, demonstrating how the team's work experience improves is more important than showing improvements in the bottom line.

Offering training and knowledge transfer

Training is so often neglected or seen as a luxury item in the project budget. But changing tools is like driving a new car with a manual transmission—a driver who has experience with automatic transmissions only is going to be very frustrated.

Your results won't be much better than the poor driver if you institute new tools and processes without training your staff. When you put together a project plan and a budget to implement new content processes, it is essential to include knowledge transfer as part of the costs. In addition to spending money on new tools, you need to show your team how to use those tools effectively.

There are a few options to consider for knowledge transfer, including:

- *Classroom training.* Generally the best way for team members to understand a new process and develop skills. Personal interaction with an instructor provides invaluable feedback. Smaller classes with no more than 10 to 12 students are better to ensure more one-on-one communication with the instructor. If you have a big team, consider splitting the group into multiple training sessions.

- *Live web-based training.* Particularly cost-effective for geographically dispersed teams. Recordings of the web sessions provide a great resource for team members who want a refresher course on the tools or who join your staff later.

- *Train the trainer.* Train one or two team members and then have them share their expertise with the rest of the group. While not as ideal as having a professional instructor teach your team, this scenario can be effective if the team members offering the training have an excellent grasp of the new process, and have the skills (especially patience!) to demonstrate what they know to their coworkers.

The benefits of knowledge transfer are two-fold: team members can ramp up on the new processes in less time (thereby more quickly achieving the cost savings that upper management likes so much), and the team members themselves gain new skills in their profession.

Differentiating between legitimate issues and recalcitrance

During periods of change, some level of resistance is a given, and not all of it is bad. Often, your staff's pushback can help identify deficiencies in the new workflow. Addressing such feedback can help you win "converts" to the new process. They can become allies who help you gain the support of other team members with lingering reservations about the changes.

No matter how well you respond to feedback, a percentage of staff will be dead set against changes on general principle. You will not win them over with explanations, training, chocolate, or anything else. The goal of a well-run change management process is to minimize the adamant opposition.

You may need to consider other assignments for such an employee: for example, maintaining legacy documentation in the old system, particularly if that employee has extensive domain knowledge. You must weigh the value that this employee provides to the organization against the drag introduced by general recalcitrance and its effect on other team members.

Using pilot projects

Responding to feedback gains you converts who can win over others in a department adopting a new workflow. These converts are also an invaluable resource for explaining the benefits of change when new content processes are implemented across a company.

This is particularly true if you start with a pilot project that affects just one department (or a smaller segment of a big department). The pilot project enables a core group to develop a process and work out any kinks before expanding the changes across the organization. The participants in the pilot then act as "evangelists" who explain process change to other employees as the new workflow is rolled out to other groups.

These evangelists can gather feedback from the new groups and work with the process implementers to address department-specific challenges and any newly identified deficiencies in the overall process. Feedback from those who were not involved in the pilot can sometimes uncover issues the primary participants missed.

Generally, employees facing changes to their workflow are more receptive to hearing another employee explain how an updated process made things better (*"To create PDF and HTML versions of content, I press two buttons. That's it!"*) than listening to management talk about process change in less specific terms. However, when employees from one department are sharing information about new content workflows with another department, they need to be diplomatic and sensitive in their approach (particularly if there is a history of tension between the departments).

It would not be wise, for example, to have employees from the tech comm department go to the marcom group and tell those employees they must use the tech comm methodology. Instead, the tech comm employees should explain the benefits the tech comm department reaped from the new workflow and then talk with the marcom group about how to apply *and adapt* the processes for marketing content. A good content strategy can accommodate differences in content while still enabling content collaboration and streamlining the development and distribution of information.

Appendix A: Creating useful information

The foundation of content strategy is useful information. Unfortunately, technical content isn't always useful. The reasons for this vary, but most often, the following factors are at work:

- *Content is an afterthought.* Technical content is seen as a necessary evil and not a strategic asset. Documentation is created at the last minute, content creators are not full members of the product teams, and the "just get something out there" mentality prevails. Authors pick up whatever tools they can—Microsoft Word, an open source wiki, web-based word processors, and so on.

- *Content solves the wrong problem.* This problem is especially common in software documentation, where authors focus on providing click-by-click instructions ("Click the Name field and type in your name") instead of useful, hard-to-find information ("Omit apostrophes and any other special characters, which result in an 'invalid name' error message").

  Note: In many cases, the critical information needed in technical content reflects product design problems. The Name example would be better addressed by creating software that either supports apostrophes or by having the software remove the characters and notifying the user ("Using OKEEFE instead of O'KEEFE").

- *Content creators have the wrong skillsets.* Content creators must be capable of understanding the product and explaining it to others.

Many product developers fail the second test. Some writers fail the first test. Effective technical communicators balance these two requirements.

Content needs to be:

- *Correct.* You cannot save technically incorrect information with snazzy formatting.

- *Relevant.* Accurate information is not enough—readers want information that addresses their specific issue.

- *Concise.* Most readers do not want to wade through huge volumes of information to find what they need.

- *Accessible.* Readers must be able to get at the information; that means, for example, providing graphics that do not use tiny type and addressing the needs of readers with varying degrees of literacy, visual acuity, computer skills, and so on.

- *Usable.* Readers must be able to find the information they want and understand it.

This appendix offers a quick overview of factors that contribute to useful content. For detailed information about developing effective technical content, refer to *Technical Writing 101: A Real-World Guide to Planning and Writing Technical Content* (scriptorium.com/books/technical-writing-101).

Writing

Technical writing breaks down complicated information into content for a specific audience. In general, technical content should be:

- Clear

- Easy to understand

- Not subject to misinterpretation

- Concise

- Easy to follow

Grammatically pristine content that merely rehashes self-evident information is not useful technical information ("Press the **Print** button to print"). Good technical content provides information end users cannot easily discern on their own, and it also offers context by explaining how one action can have effects beyond what a user may have intended ("If you activate duplex printing on a printer that does not support it, only the odd-numbered pages will print"). An author must have a deep understanding of a product to offer that level of information, so technical aptitude is as essential as good writing skills.

A detailed discussion of writing technical content is beyond the scope of this book, but some basic principles of clear technical writing are as follows:

- Know your audience

- Use active voice

- Be concise

- Use the second person ("you can…" not "the user can…")

Quality technical content is easier to understand, which means that you open up your information to readers who are:

- Lower literacy

- Reading in a non-native language

- Reading about an unfamiliar topic

Editing

Fewer and fewer tech comm departments employ full-time technical editors to review content, so many groups rely upon peer reviews among technical authors to:

- Improve the organization, tone, and consistency of technical content

- Correct spelling and grammatical errors

Some companies use software to check for grammatical errors and inconsistent terminology.

Editing can ensure that content clearly communicates the necessary information at the appropriate audience level; this sort of feedback is invaluable while developing content. Also, every error caught during the editing process is a mistake end users will not see—which translates into fewer dings to your company's reputation.

> **Note:** When a company takes a modular approach to documentation, the order in which a reviewer or editor reads topics may not follow the order of topics in a particular deliverable (book, help system, and so on). Different deliverables (and even different versions of one deliverable) can be created by including or excluding particular topics. Because of all the possible topic combinations, it's tough—if not impossible—for a reviewer to get a handle on how content flows when topics are combined for various deliverables. Therefore, in topic-based writing environments, it may make more sense for the editing process to focus more on the consistency of how information is presented.

Accessibility

The term *accessibility* has many different facets, but generally refers to making content usable by as many people as possible. The most common accessibility concern is people with visual impairments, who need the ability to use a screen reader or to magnify text. But accessibility goes beyond that:

- Graphics should have alternate text captions (which a screen reader can interpret) to accommodate people with visual impairments.

- Video should have audio that provides complete information (instead of relying on visuals on the page to convey information). For people with hearing impairments, video should also support closed captioning and transcripts.

- For people with cognitive impairments or limited language proficiency, short sentences and simple words are crucial.

- For people with physical impairments, an alternative to mouse-driven commands is needed.

Producing accessible content boils down to some simple best practices:

- Never rely on a single sense (such as vision or hearing) to convey meaning.

- Deliver well-organized content that is easy to scan.

- Create well-written content that is easy to understand.

- Include accessibility considerations in the initial content design and in the authoring process.

Content accessibility is similar to some of the considerations that are used in "universal design" by building architects. It is less inexpensive to plan for accessible content and include accessibility as a key requirement than it is to retrofit accessibility "features" onto the content later. Similarly, it's much less expensive to design a house with wide doorways, grab bars, and the like than it is to add these components later.

In some jurisdictions, providing accessible information is a legal requirement. But whether there is a mandate or not, ensuring that content is accessible opens up your information to the widest possible audience.

User experience

When applied to content, *user experience* encompasses design, organization, accessibility, and usability. Can your users quickly find the information they need? Is the process of finding and using information pleasant?

For example, if your primary end users are technicians who carry iPad tablets in the field, you should supply content that is easily accessed and viewed on those devices. A highly designed PDF file with hundreds of pages is probably not the way to go. Instead, lightweight, modular HTML with strong search and navigation capabilities is a better choice.

Understanding your audience goes beyond just how you write content— it is also a crucial factor in deciding how you distribute it.

There is an entire community of practice devoted to creating effective web sites. Good user experience needs to be part of your content strategy.

Localization

Localization is the process of modifying content (or a product) to make it usable for a new locale. Often, this includes translating the content from the source language into the language used in the locale. If, for example, you are selling in the U.S. market and want to expand into Germany, you probably need to translate your content into German. The localization process, however, includes more than just translation. For example, U.S. and German currencies are different, so you might need to change references from dollars to euros. It is possible to localize information *without* translating it—such as changing examples, currency, geographically specific information, and perhaps color schemes.

There are some standard best practices that help to minimize the cost of localization, including the following:

- Source text should use clear, concise language

- Do not use jargon, idiom, metaphors, puns, or other creative language

- Use simple words

- Use simple language structure

These best practices are appropriate for technical content, especially procedures and reference information, which need to be highly structured anyway. The readers in the source language also appreciate information that is presented consistently. For example, in providing menu instructions, choose one standard, such as **File > Open**, and use it consistently. Do not use any other variations, such as: choose **File**, then **Open**; select **Open** from the **File** menu; from the **File** menu, select **Open**; or click **File**, **Open**.

Following these best practices will help to reduce the overall cost of localization. Translators can more easily understand the content and deliver it in other languages and, importantly, they can use translation memory tools that recognize patterns and support some automation in translation.

The cost of efficiency

Optimizing content for efficient translation means letting go of interesting word choices, pop culture references, and other creative communication approaches. The result is bland sentences, which is acceptable for most technical content. But if your communication strategy requires you to make an emotional connection with the reader, you may need to use more creative language. Be aware that these approaches make translation more expensive—it takes longer for the translators to develop an equivalent phrasing in each target language. For example, the expression "once in a blue moon" cannot be translated literally because "blue moon" doesn't have a special meaning outside of English. Instead, the translator must think of a metaphor that means "very rarely" in the target language and that fits the tone of the message.

Visual communication

Translating text is challenging, but graphics, video, and other visual components are even more difficult. You can create graphics that are culturally neutral and appropriate for multiple languages, but there are myriad challenges, including the following:

- The meaning of a color varies across cultures.

- Images of people are not appropriate in some cultures. In other cultures, images of women are not acceptable.

- Visual representations of hands or feet are not appropriate in some cultures.

- It is difficult to design icons that are universally understood.

- Visual metaphors or puns may not work in translation. (For example, the idea of an owl is a symbol of wisdom in the United States and parts of Europe. But in other cultures, owls are associated with death and witchcraft.)

Automating translation

For technical content, translators generally charge by the word. At 10–25 cents per word (per language), translation expenses rise quickly. For example, a 100-page document would have approximately 25,000 words, so translation would be at least $3,000 per language. (This quick estimate does not factor in project management, graphics, or anything else that might complicate the process.)

The idea of automated translation is deeply appealing—computers can do nearly everything else, so why not this?

There are two major categories of automated translation:

- *Machine translation.* This refers to a process where the source language is fed into the system and the target language is generated as output. The quality of the translation is not very good, but it is usually sufficient for the reader to understand the general meaning of the original text.

- *Computer-assisted translation.* Translation memory (TM) systems contain a database of previously translated language pairs. When a new translation project is loaded into the system, the TM is scanned for matches. Thus, if you have already translated a sentence ("The database is read-only") into multiple languages, and that sentence occurs again in new content, the TM system will recognize that there is a match and translate that sentence for you. When you load in a sentence that is similar to previously translated content ("The file is read-only"), the system may recognize this as a "fuzzy match" and provide you with a proposed translation. Human intervention is then required to validate the fuzzy matches and ensure that the final document makes sense in context.

Computer-assisted translation is the current standard and should be used in any translation effort so that the translation is stored in a reusable manner for future use. Most organizations that use professional translation companies have translation memory in place.

Many organizations today are using "in-house" translation. In some cases, they have a professional translation team, but often, the translation process involves shipping the content to in-country employees (often, sales/marketing or engineering staff) for translation. This approach is acceptable for a one-time translation of a small amount of information, but any company with serious global operations needs to invest in a professional translation workflow.

Optimizing your localization process

There are several opportunities to ensure that localization is as efficient as possible.

Table 26: Opportunities for efficient localization

Stage	Best practice
Content development	Create clear, simple, concise text. Limit text in graphics, and put text into an editable layer. Use legends rather than embedded text for graphics.
Formatting	Use predictable structure. Use styles or structured authoring. Do not use embedded formatting.
Delivery	Include a language flag on content so the system can deliver content to readers in the language they want.
Localization	Use professional translators. Use translation memory systems. Work with localization professionals to identify potential problems. Look for opportunities to change source content instead of requiring changes in each target language.

Culture

Understanding the impact of culture is a key part of creating useful, usable information. First, consider your audience:

- Does your audience have a common culture?

- What sorts of things do your audience members have in common?

- How should those common factors affect your design?

- What is the best way to support your audience?

- Is your audience change-averse? What is the best way to introduce changes?

This is basic audience analysis; nothing new to see here. There are, however, other cultural factors that you need to consider. These include the culture of your content creators and the corporate culture.

Corporate culture affects what information products you can create. For example, a risk-averse corporation with a penchant for secrecy is going to have a tough time with publishing user documentation for the world to see. A startup company with limited funds and aggressive deadlines may want to lean on the user community to produce content—even though, as a startup, there might not be a user community yet!

The risks associated with using your product matter to the culture of your audience. Gamers have different expectations for game documentation than surgeons do for medical devices. Documenting software used in a medical setting requires a different attitude than documenting a game.

Beyond the corporate culture and audience, you also need to consider the culture of the locales where you are active. Local culture should affect your choice of branding, colors, presentation, and idiom.

User-generated content

If you want your user community to produce content (as opposed to commenting on content you create), you need to give them the tools to do so. One common approach is to create a wiki and build an initial structure that reflects what you think people will want to write about. Once the actual users get involved, they will probably change the organization, but prepopulating the wiki with some information and a proposed organization helps encourage participation.

Adding a social layer, which lets readers comment on information and share it via social media channels, is technically not very difficult. The challenges lie on the policy side:

- What sort of comments are allowed?

- Will you moderate comments? If so, what is the moderation policy?

- How will you handle critical comments?

- How will you handle comments that are technically incorrect?

Versioning presents another major challenge. Let's say your product is in version 1, and your users have commented extensively on the version 1 content. When you update content for version 2, what do you do with the comments? Do you keep them or remove them? What if you corrected the content based on a comment? Can you now delete the comment?

Finally, you need to think about reputation issues. Within your user community, each participant will have a reputation, which may be positive (expert knowledge, always helpful), negative (sarcastic, rude, and often inaccurate), neutral (lurker, never posted before), or something more nuanced. For example, a product manager might have a reputation for glossing over his own product's deficiencies while highlighting problems with competitive products. However, this manager might be a reliable source on the overall industry. You cannot control the reputations directly, as David Lankes writes:[26]

> Any system that seeks to either impose an authority view of credibility, or that seeks to change behavior must now do it with the understanding that users can simply bypass these attempts and create counter structures. Furthermore, these alternative credibility structures can have a global reach and build communities of like minds across divisions of geography, race, gender, age, and other demarcations.

If you attempt to lock down and control the user community beyond what *the users* consider reasonable, you may end up with no community at all.

---

[26] "Credibility on the internet: Shifting from authority to reliability," *Journal of Documentation,* Vol. 64, No. 5, 2008, R. David Lankes, quartz.syr.edu/rdlankes/Publications/Journals/credibilityontheinternet.pdf, accessed March 29, 2012

Index

Numerics

A

B

C

Colophon

This book was produced from DITA sources using the DITA Open Toolkit Release 1.6.1 and the Antenna House Formatter v6.0. Our spdf2 plugin written by Simon Bate is based on the Open Toolkit pdf2 plugin and adds a number of enhancements. These include:

- Detailed control of common layout elements

- Treatment of <section> elements as subordinate topics

- Correct handling of <part> elements in bookmaps and supporting unique images for each part

- Improved appearance of admonitions (note, caution, warning, and so on)

- Repeating table titles (with "continued") on page breaks

- Improved widow and orphan handling in tables

- Extensive table customization

- Options for formatting definition lists (<dl>) as a table or as a list

- Support for two-column output of simple lists

- Automatic generation of cover and copyright pages from bookmap metadata

CPSIA information can be obtained at www.ICGtesting.com
Printed in the USA
LVOW10s0255060214

372439LV00004B/505/P